Praise for
Gerald Weissmann

"Weissmann is Lewis Thomas's heir." —**Robert Coles**

"How I envy the reader coming upon Dr. Weissmann's elegant, entertaining essays for the first time!" —**Jonas Salk**

"Dr. Weissmann's juggling with the balls of global politics, biology, medicine, and culture in the framework of history is breathtaking."
—**Bengt Samuelsson**, Nobel laureate and former chairman of the Nobel Foundation

"The premier essayist of our time, Gerald Weissmann writes with grace and style." —**Richard Selzer**

"An absolutely first-rate writer." —**Kurt Vonnegut**

"As a belles-letterist, Weissmann is the inheritor of the late Lewis Thomas. . . . Like Thomas, he's a gifted researcher and clinician who writes beautifully. Unlike Thomas, he is an original and indefatigable social historian as well." —***Boston Globe***

"He writes as a doctor, a medical scientist, a knowing lover of art and literature and a modern liberal skeptic. But more than anything else, Weissmann writes as a passionate and wise reader." —***New Republic***

"Weissmann is a master of the essay form. His witty and elegant prose makes the toughest subject matter not only accessible but entertaining."
—***Barnes and Noble Review***

"[Weissmann] is a Renaissance Man. . . . He'll stretch your mind's hamstrings." —***Christian Science Monitor***

"[Weissmann's essays] intertwine the profound connections of science and art in the context of our modern era . . . to illuminate the ongoing challenges scientists face in dealing with scrutiny and criticism, from colleagues and from our broader society." —***Science***

"Erudite, engaging, and accessible." —***Library Journal***

"Essays that brim with knowledge and bubble with attitude."
—***Kirkus Reviews***

"Weissmann models his work after that of his mentor, Lewis Thomas. . . . His ideas . . . are every bit as important." —***Publishers Weekly***

THE FEVERS OF REASON

Also by Gerald Weissmann:

The Woods Hole Cantata

They All Laughed at Christopher Columbus

The Doctor with Two Heads

Democracy and DNA

The Doctor Dilemma

Darwin's Audubon

The Year of the Genome

Galileo's Gout

Mortal and Immortal DNA

Epigenetics in the Age of Twitter

THE FEVERS OF REASON

New and Selected Essays

Gerald Weissmann

Bellevue Literary Press
New York

First published in the United States in 2018 by Bellevue Literary Press, New York

For information, contact:
Bellevue Literary Press
NYU School of Medicine
550 First Avenue
OBV A612
New York, NY 10016

Library of Congress Cataloging-in-Publication Data

Names: Weissmann, Gerald, author.
Title: The fevers of reason : new and selected essays / Gerald Weissmann.
Description: First edition. | New York : Bellevue Literary Press, 2018. |
Includes bibliographical references and index.
Identifiers: LCCN 2017039245 (print) | LCCN 2017043326 (ebook) |
ISBN 9781942658337 (e-book) | ISBN 9781942658320 (pbk.)
Subjects: LCSH: Science—Miscellanea. | MESH: Science—history | Philosophy,
Medical | History, 20th Century | History, 21st Century | Essays | Collected Works
Classification: LCC Q173 (ebook) | LCC Q173 .W4424 2018 (print) |
NLM Q 126.8 | DDC 500—dc23
LC record available at https://lccn.loc.gov/2017039245

Bellevue Literary Press would like to thank all its generous donors—individuals and foundations—for their support.

This publication is made possible by the New York State Council on the Arts with the support of Governor Andrew Cuomo and the New York State Legislature.

This project is supported in part by an award from the National Endowment for the Arts.

Book design and composition by Mulberry Tree Press, Inc.

Manufactured in the United States of America.

First Edition

1 3 5 7 9 8 6 4 2

paperback ISBN: 978-1-942658-32-0

ebook ISBN: 978-1-942658-33-7

My love is as a fever, longing still
For that which longer nurseth the disease,
Feeding on that which doth preserve the ill,
The uncertain sickly appetite to please.
My reason, the physician to my love,
Angry that his prescriptions are not kept,
Hath left me and I desperate now approve
Desire is death, which physic did except.
Past cure I am, now reason is past care,
And frantic-mad with evermore unrest;
My thoughts and my discourse as madmen's are,
At random from the truth vainly express'd;
For I have sworn thee fair and thought thee bright,
Who art as black as hell, as dark as night.

—William Shakespeare (1609)

Le jeunesse est une ivresse continuelle; c'est la fièvre de la raison.

—Duc de La Rochefoucauld (1665)

To Ann

toujours toujours là pour moi

Contents

Prefatory Note

> Ideas and products and messages and behaviors spread just like viruses do.
>
> —Malcolm Gladwell, *The Tipping Point* (2000)

NOT ONLY LOVE, as the Bard tells us, or youth, as the Duke suggests, produces fevers of reason. We've learned that the fevers of Zika and Ebola can sear the mind; we've also learned that reason becomes toast when presidential Tweets go viral at dawn. Fevers of reason require treatment based on facts not fancy, brains not bravado.

Happily enough, messages of cool reason can also go viral, and at their best, inform and command. That's especially true of scientific papers that introduce notions like the helical structure of DNA. To become viable, and go viral in turn, their progeny must survive the birth pangs of test, retest, and peer review. When a tested notion reaches adolescence, we call it a hypothesis (DNA makes RNA makes protein). When a hypothesis reaches maturity it becomes a theory (relativity) and, with time, becomes a law (gravity).

No such direct path for an essay. While the word comes from the French *essai,* a "test, trial, or experiment," essays don't require independent proof. The essay form may have been set by Montaigne's musing from the terrace of his château, but those boundless vistas are long gone. Essays today, mine included, are written in the realm of a viral Internet, where imaginary gardens of fancy have yielded to a thicket of facts and "alternative facts" through which reason is the surest guide.

In the first four sections of this book, the essays deal with four themes to which I've returned over the years. "Going Viral" connects a 2016 Woods

Hole lecture on gene splicing to a fictional discovery made in Sinclair Lewis's 1925 novel *Arrowsmith*: a virus that cures the plague. We also find a link between the Ebola epidemic in West Africa and gun violence in the United States, both of which have gone viral. We trace Zika virus and the drug thalidomide to fetal abnormalities, forms of which have not deterred their victims from fame or infamy. Figuring out how viruses turned purple petunias white led to a Nobel prize and an inscription on a monument on Columbus Avenue in New York City, while a disease of British royalty (lupus) directly affected the history of health care in the United States.

"Science Fictions" lays out histories of the age-long contest between experimental reason and febrile beliefs such as "animal magnetism," "homeopathy," "distant healing," and "intelligent design." We meet Dr. Oliver Wendell Holmes and Anton Mesmer, Edgar Allan Poe and Dr. Arthur Conan Doyle. We conclude that truth is no stranger to fiction.

In "Two for the Road," we remember lifelong partnerships between two people, married or not, unions that have given us oxygen (the Lavoisiers), radium (the Curies), women in medicine (the Blackwells), "America the Beautiful" and the New Deal (Bates and Comans), and the neatest prose written by a member of the James family (Alice James and her Katharine).

The fourth section, "Beside the Golden Door," considers contributions made to American science by immigrants who passed by that Lady lifting her lamp. I write about Albert Einstein and Currier McEwen, then dean of the NYU Medical School, and about Richard Dawkins, who returned to Oxford from the United States with a Gene McCarthy sticker on his Ford. I contrast the experiences of scientists like Jan Vilcek and Roslyn Yalow, who became Americans to escape racial prejudice in Europe, with that of Percy Julian, a black American scientist who learned chemistry in Vienna and returned to face racial prejudice at home.

The last chapter is a personal memoir dedicated to my mentor in the lands of reason, Lewis Thomas, a polymath who merged arts and science in all his works. He was also a mean wit who enjoyed H. L. Mencken's quip "Martinis are the only American invention as perfect as a sonnet." I'd say that Thomas's first book of essays, *The Lives of a Cell*, is a martini—with a twist. Cheers!

Going Viral

1.

Arrowsmith and CRISPR at the Marine Biological Laboratory

> As the audience flows out of the auditorium [after] the Friday Evening Lecture, the MBL's weekly grand occasion when the guest lecturers from around the world turn up to present their most stunning pieces of science, there is the same jubilant descant . . . half shout, half song made by confluent, simultaneously raised human voices explaining things to each other.
>
> —Lewis Thomas, *The MBL* (1972)

On one of those grand occasions, Jennifer Doudna of Berkeley presented the latest news of CRISPR to a packed auditorium at the Marine Biological Laboratory. Her audience was the usual mixed crowd one finds at the Friday Evening Lectures in Woods Hole. I spotted working scientists, grad students, lab assistants, undergraduates, a score of academicians, a Nobel laureate or two, attentive families and friends, hailing from all corners of the globe. The dress code ranged from country jeans to khakis, bike gear to saris, backpacks to bow ties.

After the smartphones were turned off and the tablets stowed, Doudna proceeded to hold the audience in thrall for an hour with bulletins from the front lines of the war against error—in the gene.

Doudna began by spelling out the acronym CRISPR, explaining that it stands for "clustered regularly interspaced short palindromic repeats" in the DNA of bacteria. When viruses called bacteriophages infect a bacterium, the CRISPR system filches DNA from the phage virus and inserts it into its own DNA. That genetic memory of the encounter will be passed to its progeny. In 2005, French scientists (Pourcel et al.) studying genes of a bacterium (*Yersinia pestis*) that caused plague in 1960s Indochina, found remnants of bacteriophage DNA at CRISPR spacers in the plague bacilli. They proposed that the locus "may represent a memory of past genetic aggressions." With its neighboring *cas* (CRISPR-associated system) genes, CRISPR works in bacteria like a smallpox vaccination in humans, providing adaptive immunity to a virus.

CRISPR generates a unique set of RNAs that guide Cas proteins directly to the DNA of any future phage aggressor. The neatest of the Cas proteins is Cas9, from a streptococcus; it's a DNA-cleaving enzyme. But progress with the Cas9 system for gene editing had been hampered because it required two different guide RNAs. Doudna detailed the remarkable contribution she and Emanuelle Charpentier made to the field in 2012, as described in Jinek et al. They engineered *single,* specific RNAs that could guide the Cas protein to cleave DNA of any species at any given site, permitting normal DNA repair by the cell's built-in machinery.

By excising an unwanted gene and replacing it with a desired substitute, they'd waved the magic wand of genetic engineering. Doudna went on to show dazzling animations in which twists in RNA, tweaks of DNA, and acrobatics of the Cas proteins accomplished the task. She gave examples of how the method had already been applied to modify the color of mushrooms and to repair the faulty genes in models of muscular dystrophy.

Doudna closed by warning that perhaps "the science is going too fast." While CRISPR technology can erase crippling misprints in our genes, there remain ethical roadblocks to extending the method to cells of the human germ line. At present, she argued, CRISPR should be kept away from human sperm and eggs until there is a general consensus as to how, when—and if. The audience was clearly in accord and showed their agreement by waves of applause at the end of the talk.

As the audience moved out over the steps of the auditorium, one heard—as Thomas put it—the customary descant of "confluent, simultaneously raised human voices explaining things to each other."

A senior scientist enthused, "Great stuff! No wonder her paper went viral! But, on the other hand . . ."

A course instructor interrupted, "You know, that egg slide of Doudna's looked like the old MBL physiology experiment: you get sea urchin eggs to divide by changing salts rather than adding sperm."

A grad student chimed in, "That was Jacques Loeb, the guy who got famous for parthenogenesis!"

"Wasn't he the old prof with the accent in *Arrowsmith*?" a jean-clad fan of oldies asked.

That one clicked: "Yeah sure, Loeb *was* Dr. Gottlieb in *Arrowsmith*, the book *and* the movie." . . . "I think the young doc tried to use phage to cure the plague." . . . "Great movie, but the wife died."

As the group broke up and traipsed along to the evening reception, it seemed to me I had heard that song before. Sure enough, guided by the rich archives of the MBL library and my tattered copies of Loeb's *The Mechanistic Conception of Life* (1912) and Sinclair Lewis's *Arrowmith* (1925), I came to the conclusion that Doudna's lecture couldn't have been given at a location more appropriate—nor on a topic more likely to go viral.

The Friday Evening Lectures at the MBL in Wood's Holl, as it was then called, descend from a series of public lectures that began in 1889, the year after the lab was founded. And over the next decade, Jacques Loeb (1859–1924) gave several of these, published in *Biological Lectures at MBL Wood's Holl.* Based on his summer experiments at the MBL on sea urchin sperm and eggs, in 1909 he issued a challenge:

> Whoever claims to have succeeded in making living matter from inanimate will have to prove that he has succeeded in producing nuclear material which acts as a ferment for its own synthesis and thus reproduces itself.

Doudna and the other "heroes of CRISPR," as geneticist Eric Lander called them, met that challenge just over a century later. The CRISPR-Cas method uses "inanimate RNA" and protein to snip out the old "nuclear material" (DNA) to add the new which can "act as ferment" for its own reproduction. Going viral, we might say.

CRISPR-Cas HAS ALSO GONE VIRAL in the scientific literature. According to the website Web of Science, the topic of CRISPR-Cas went from a few dozen citations in 2005 to over ten thousand on the eve of Doudna's lecture. Her seminal 2012 paper with Charpentier has itself been cited over fifteen hundred times. (To put this into current pop perspective: one notes that one tweet by Kim Kardashian has reproduced itself 137,369 times.)

As for Jacques Loeb, he went as far as one could in the pre-viral days. His papers were cited fifteen hundred times in his lifetime, thrice the number of his contemporary, Paul Ehrlich. He was also a figure in the press: "Loeb Tells of Artificial Life," proclaimed the *Chicago Daily Tribune* in 1900. In1907, under the headline "Believes Germ of Life Will Be Discovered," the *San Francisco Call* reported that "Professor Jacques Loeb . . . in a bulletin issued today from the office of the president, makes the statement that he believes the germ of life can be discovered, provided the chemical reactions surrounding the process of fertilization, are investigated."

Loeb's public persona also went viral in his role as the fictional Dr. Gottlieb, the mentor in *Arrowsmith*. The novel by Sinclair Lewis won a Pulitzer Prize, and the film was nominated for an Oscar. Lewis described how Gottlieb and Martin Arrowsmith (Ronald Colman on screen) not only discovered bacteriophage, but used it to stem an outbreak of plague (*Y. pestis*) on a tropical island.

BACTERIOPHAGE AND THE PLAGUE may have been brought to modern attention in 2005 when French scientists found phage-like DNA sequences at the CRISPR locus of plague bacilli, but Martin Arrowsmith got there first! His story was patched together from Paul de Kruif's memories of work at Rockefeller Institute (now Rockefeller University) and Sinclair Lewis's progressive politics. De Kruif and Lewis made a joint field trip to the West Indies, which gave Lewis the setting for an epidemic of plague in a colonial context; it also set the stage for an exercise in bioethics.

Plucked from an earnest career as a Midwest clinician, Martin Arrowsmith is taken up by the McGurk (read Rockefeller) Institute, where he stumbles on filterable "Factor X," which seems to kill bacteria, staph at first. But then, Gottlieb (Loeb) steps in:

> "Why have you not planned to propagate it on dead staph? That is most important of all."
>
> "Why?" Gottlieb flew instantly to the heart of the jungle in which Martin had struggled for many days: "Because that will show whether you are dealing with a living virus."

Martin continues the work and finds not only that this virus kills several kinds of bacteria but also that he can make batches of the purified material to test in infected animals. He's ready to publish his work, when Gottlieb brings bad news—in a German accent:

> "Vell. Anyvay. D'Hérelle of the Pasteur Institute has just now published in the Comptes Rendus, Académie des Sciences, a report—it is your X Principle, absolute. Only he calls it 'bacteriophage.' . . . Bacteriophage, the Frenchman calls it. Too long. Better just call it *phage*."

Martin goes on to further basic, laborious studies on phage and, lo and behold, he discovers a century before Doudna's work that his phage "can cause mutations in bacterial species." He plugs on but is soon confronted by the director of the McGurk Institute, Dr. Tubbs (read Simon Flexner, the first director of the Rockefeller Institute), who faults Martin for not making his discoveries go viral and makes a suggestion:

> "D'Hérelle's discovery hasn't aroused the popular interest I thought it would. If he'd only been here with us, I'd have seen to it that he got the proper attention. Practically no newspaper

> comment at all. . . . I think it may now be time for you to use phage in practical healing. I want you to experiment with phage in pneumonia, plague, perhaps typhoid."

As expected, Martin gets phage to cure plague in experimental animals—and people come next. There's plague in the Caribbean, on the quarantined island of "St. Hubert," a colony beset by rats and shiftless administrators. Arrowsmith heads there with a team of microbe hunters and volunteers, including his faithful wife, Leora (Helen Hayes in the film). They'll try to stem the epidemic with a batch of *Yersinia* phage that's cured plague in rats.

The team faces personal tragedy and ethical conflict. For any new treatment, science demands controls, but can an investigator willfully deprive anyone of lifesaving measures? Does protocol trump empathy? *Arrowsmith*, book and film, throws a curveball: Martin's beloved Leora dies from the plague and a depressed Martin stands down from his task. Others prevail, and *Yersinia* phage is finally given to one and all. Since this is fiction, the effort succeeds and the plague is quenched.

A tear-filled Martin returns to New York, where the McGurk honchos are in bliss: the newspapers have gone viral with news of the phage and the plague. The director congratulates Martin:

> "You have done what few other men living could do, both established the value of bacteriophage in plague by tests on a large scale, and saved most of the unfortunate population."

In Martin Arrowsmith's fictional experiments with phage, empathy ran ahead of clinical science. In the case of CRISPR-Cas9, bacterial immunity to phage, the opposite may be true. Doudna and others, David Baltimore among them, have argued that since CRISPR can be used to hand-pick the DNA of our gonads, human evolution could be played out at the lab bench. If the science has moved too fast for the ethics, perhaps the time has come to stop and think about Lionel Trilling's warning:

> The apparatus of the researcher's bench is not nature itself, but an artificial . . . contrivance much like the novelist's plot, which is devised to foster or force a fact into being.

The plot of CRISPR is before us, but we're only into the first chapter of this book—and we're not talking fiction.

2.

Ebola and the Cabinet of Dr. Proust

> The death of millions of people whom we do not know barely touches us, and almost less unpleasantly than a current of air.
>
> —Marcel Proust, *Remembrance of Things Past* (ca. 1916)

> *Cordon sanitaire*—we call it a *cordon* when someone who presents any symptoms of that disease is transferred, if their condition permits, to a hospital or to an equivalent place designated by the local authorities.
>
> —Adrien Proust, "Essay on International Hygiene" (1873)

> We believe these new measures [taking temperatures and administering questionnaires at JFK and other airports] will further protect the health of Americans, understanding that nothing we can do will get us to absolute zero risk until we end the Ebola epidemic in West Africa.
>
> —Thomas R. Frieden, Centers for Disease Control and Prevention (October 8, 2014)

WE'VE LEARNED FROM THE EBOLA OUTBREAK of 2014 that to stop pandemics, we must regulate the transport of host and virus alike. Public transmission and cellular invasion depend on the unimpeded traffic of people across borders and of viruses across cell membranes. Barring the traffic of people across the Atlantic from Liberia would certainly have prevented the first four cases of Ebola in the United States. Blocking the traffic of the virus in human cells would have prevented thousands of cases worldwide. The Ebola epidemic in West Africa presented a major challenge not only to our public health system but also to our capacity to develop antiviral drugs. Thanks to cell and molecular biology, we already understand some of the critical pathways of virus entry and replication. We've also learned that these conduits can be blocked by novel drugs such as dynasore [*sic*] and Dyngo (no kidding) that are slowly coming to the clinic. To everyone's surprise, papers by Harper et al. and Masaike et al. showed that bisphosphonates, drugs widely used to treat osteoporosis, also inhibit the traffic of Ebola virus in cells.

So now that we have the tools available—dynasore, vaccines, adenosine analogues, inhibitors of RNA dependent-RNA polymerase, and so on (as shown by Oestereich et al.)—isn't it time to launch a major effort, like those against smallpox, polio, and AIDS, to bring the benchwork of pharmacology to the bedsides of West Africa?

IF DRUGS OR VACCINES HAVEN'T BEEN DEVELOPED for an infectious disease, there are only two time-tested ways to quell an epidemic: quarantine and sanitation. In the case of Ebola, the first step would have been to stop reliance on thermometers and telephones to keep infected victims from transmitting the disease. Temperatures vary, Tylenol can mask fever, and reliable information via telephones depends on who's asking and who answers. Those protocols didn't work in Dallas in the case of the first victim, the unfortunate Thomas Eric Duncan. Also fallible was the CDC's policy of voluntary self-monitoring by people exposed to active Ebola victims: ask the many passengers between Dallas and Cleveland. The third American Ebola victim, Amber Joy Vinson, got permission for flights between Dallas and Ohio "because her elevated temperature of 99.5 degrees was below the no-fly threshold of 100.4 degrees," the *New York Post* reported. Once it became known that Ms. Vinson had been involved in Mr. Duncan's care, Dr. Frieden of the CDC confessed that "she should not have traveled on a commercial airline"—let alone take a taxi or visit a popular bridal shop. In contrast, quarantine of Mr. Duncan's immediate family seemed to have worked.

The CDC and the White House presented cogent arguments against a strict ban on flights from Liberia, Guinea, and Sierra Leone, arguing that such a ban would wreck the fragile economies of the area. "Trying to seal off an entire region of the world—if that were even possible—could actually make the situation worse," said President Barack Obama. A ban would also have curtailed free passage of sanitary supplies and aid workers. Many public health experts agreed and warned that the threat of Ebola to the United States and Europe would not end until its exponential spread was stopped on the ground. They're right to have worried. The early speed and extent of the 2014 outbreak in West Africa certainly dwarfed all earlier ones, and according to the World Health Organization, that outbreak was the worst ever, with 28,639 confirmed cases and 11,369 deaths by March 13, 2016. That toll and that extent are the best reasons why—next time—a travel ban, with visa restrictions and monitored quarantine, should be in place until WHO criteria for an end to the outbreak are met. That would be the quarantine part of the equation.

What about sanitation? Well, I'm afraid that all the noble efforts of Médicins Sans Frontières (Doctors Without Borders) or U.S. Army engineers were unable to produce a timely change in the sanitary culture of West Africa. Among expected obstacles, they faced unfamiliar ritual bathing and burial practices such as those that had sparked earlier Ebola outbreaks. Ebola, which is endemic in African fruit bats, first appeared in 1976 in rural Sudan and the Democratic Republic of the Congo (formerly known as Zaire): 284 were infected and 117 died. By 2000 it was Uganda's turn, with 425 cases and 224 deaths. The index case in Uganda was that of Esther Awete, a villager who died in her mud hut on September17, 2000, after several days of fever and pain. Seven of her relatives also died after they had ritually bathed Awete's corpse and washed their hands in a communal basin as a sign of communion with the dead. Ritual bathing remains common in West Africa today: three months into the 2014 epidemic, *The Economist* cited a WHO study reporting that "60% of all cases in Guinea were linked to traditional burial practices that involve touching, washing or kissing the body." Not only in Guinea: thanks to gravesite infection, Ebola remained endemic in the Democratic Republic of the Congo, with 66 cases and 49 deaths in 2014. In September 2014, WHO had called for a 70–70–60 target plan aimed at isolating 70 percent of suspected new cases of Ebola, a safe burial of 70 percent without the risk of infecting others, all within the next 60 days—and no new cases by January 1, 2015. The goal was not achieved until January 15, 2016! That's an unanswered call for quarantine *and* sanitation.

There are other factors in play. Until the 1990s, Ebola afflicted sparsely populated areas of the continent, but in 2014, the disease ran wild in Monrovia (Liberia), Freetown (Sierra Leone), and Conakry (Guinea), capitals with populations of over one million each. These cities, which retain many neighborhoods lacking clean water and adequate sewage, have also suffered from civil war and bloody coups. In Liberia, where UN peacekeepers remained until 2013, 14 years of civil war killed 200,000 of its citizens. Helene Cooper reported in the *New York Times,* "The war produced mad generals who led ritual sacrifices of children before going into battle, naked except for shoes and a gun." Sadly, there are also public health problems at the beginning of life. According to a CIA World Factbook, even before Ebola the infant mortality rate stood at 69/1000 in Liberia, 73/1000 in Sierra Leone, and 91/1000 in Guinea (versus 6/1000 in the United States).

That's why it will be a long time before we can completely revamp the local conditions that permitted Ebola to spread. Dr. Barry R. Bloom, a specialist in infectious diseases, told the *New York Times* that, in the big picture, "the most important thing that can be done to protect Americans from Ebola is controlling Ebola in West Africa." I agree, but in epidemic times we're never in the big picture, and the virus remains a latent threat which, like Zika, can arrive on the next plane.

Supporters of the original CDC protocols argued that thermometers, questionnaires, and telephones are more humane than the "medieval" solution of official quarantine. But a more rigorous model has been around for a while. Twenty-first-century Monrovia and Freetown could take lessons from nineteenth-century New York and Paris. It took the better part of that century, but the two capitals overcame five lethal epidemics of Asiatic cholera, as it was known, despite polluted water, civil war, and urban grunge. In good part this was because enlightened sanitarians came up with solutions like Croton Reservoir in New York and Baron Haussmann's roadways and sewage systems in Paris, as I described in "Cholera at the Harvey."

But sanitation alone was not the answer. The measures in New York followed the principles of the French *cordon sanitaire*: those showing signs of the disease were "taken to a hospital or to an equivalent place designated by the local authorities," according to Adrien Proust in his "Essay on International Hygiene." The places in New York designated by local authorities included quarantine ships in the East River near Bellevue Hospital and quarters at Castle Garden at the Battery, which from 1820 until 1892 was the entry point for immigrants. Inspectors washed its walks and walls with carbolic acid. Ellis Island, which opened on January 1, 1892, was equipped with larger processing, hospital, and quarantine facilities. The following September, in response to the last major cholera outbreak in Europe, President William Henry Harrison approved the last major quarantine order for New York: "no vessel from any foreign port carrying immigrants shall be admitted to enter at any port of the United States until said vessel shall have undergone a quarantine detention of twenty days . . . " —that's Ebola-level timing. A congressional act followed, but soon enough, when cholera no longer threatened from overseas, the quarantine order was lifted, never again to be used by an American president.

Opponents of quarantines against Ebola in 2014 believed that a visa-based travel ban on people from West Africa had racial overtones. We've heard that objection before. In 1892, when New York was faced with both cholera and an influx of Eastern European Jews, a prominent government official complained to the *New York Times* that "Europe is showing no anxiety to keep cholera away from us. Why should the United States accept her miserable paupers anyhow? In my opinion the President [should] ask Congress to absolutely prohibit immigration for the present." Some advocates of travel bans today may indeed be objecting to West Africans coming into our country, whatever the circumstances. I'd simply view these bans as mistaken versions of the *cordons sanitaires* that effectively stopped epidemics of cholera in the nineteenth century. We're not talking about thirty-foot walls!

The champion of the *cordon sanitaire* was Dr. Adrien Proust (1834–1903), father of Marcel Proust. Six pandemics of Asiatic cholera ravaged Europe and spread to the United States in the nineteenth century. The disease, which originated in Bengal, reached the West by land and sea; its greatest damage was wrought at times of civil or imperial war. In France, cholera reached epidemic proportions during periods of political strife in 1830–1845 and during the German siege of Paris in 1870. In 1866, Adrien Proust, clinic chief at the Hôpital de la Charité represented France at an international sanitary conference at Constantinople. Proust, who held the official title of Inspecteur Général des Sanitaires Internationaux, persuaded both European and Ottoman officials to agree that the disease was carried by contaminated water: "Water would seem, according to the observations made principally in England by Dr Snow . . . to contribute, in certain circumstances, to the development of cholera in a locality." To map the itineraries of earlier epidemics, Proust trekked to Persia, Mecca, Turkey, and Egypt. Since he was able to trace the origin of each Parisian epidemic to the Middle East, Proust pleaded repeatedly that vessels with disease on board be prevented from traversing the newly dug Suez Canal. Egypt was the key! In *La defense de l'Europe contre le choléra* he declared, "We must absolutely close the Suez Canal, to all vessels, of whatever nationality, with cholera on board or with recent exposure." Not surprisingly, Ferdinand de Lesseps, president of the Suez Canal Company, objected strongly. (Does the airline industry today oppose a flight ban?) At the Academy of Sciences, de Lesseps opposed strict cordons as "futile and inconsistent with current enlightened opinion, which now held that emanations from local miasma" were responsible for contagious spread of disease (the story is told in LaVerne Kuhnke's *Lives at Risk*).. De Lesseps and Company won out, and sure enough, the fifth cholera pandemic of 1884 arrived in France carried on ships that passed through the canal. Eventually, *cordons sanitaires* were set up as a second line of defense around Toulon and Marseilles to prevent spread of cholera to the rest of France. *Cordons* around Marseilles were so troublesome that the English gentry transferred their winter watering holes to the French Riviera: Cannes, Nice, and Villefranche became the holiday paradises of the next century.

The state-enforced *cordons sanitaires* saved Toulon and Marseilles from outbreaks of cholera. How about today? A *cordon sanitaire* had, by the fall of 2014, effectively prevented Ebola from crossing the borders of a score of African nations, while British Airways and Air France had curtailed flights. Senegal had closed its borders and suspended air flights, and Nigeria's airline canceled flights after eleven cases of Ebola were tracked to an American passenger from Liberia. Sure enough: as of October 23, 2014, WHO had already declared these two countries free of Ebola. Meanwhile, in Dallas, Mr. Duncan's family was free of Ebola after twenty-one days of enforced quarantine, and the two nurses he infected were convalescing after treatment in the most modern isolation units. That's the sort of good news a *cordon sanitaire* can bring.

There's also good news from the lab. Basic science works: dynamin, a mechano-chemical enzyme responsible for shuttling cargo in cells (Ahearn et al.), was discovered in 1989 by Howard Shpetner and Richard Vallee, then at the Worcester Foundation (now part of the University of Massachusetts). It turned out that dynamin can choreograph the pathway of viruses in cells. Dynamin not only hobbles the sticking of the virus to the cell's surface, but also promotes uptake of virus-filled vesicles from cell membranes. Dynamin opens the door to Ebola virus, as described in Aleksandrowicz et al., and helps unload the backpacks. Happily, we've moved closer from the lab to the clinic: in animal experiments, dynamin inhibitors such as dynasore already have been seen to stop the virus (Macia et al.). Better yet, the newest dynamin inhibitors (the bisphosphonates used to prevent osteoporosis) can already be found on many a woman's medicine shelf.

Has the time come to use osteoporosis drugs for Ebola? To quote Mary McCarthy's Pokey, "Who would have thunk it?"

3.

Zika, Kale, and Calligraphy: Ricky Jay and Matthias Buchinger

Phocomelia (from Gr. *phōkē* 'seal' + *melos* 'limb.') = limbs reduced to a very small volume and hidden under the integument like seals'.

—Isidore Geoffroy Sainte-Hilaire (1836)

Dr. Schuler-Faccini [is] now focused on [describing] the specific brain abnormalities of [Zika-]affected babies, as well as other associated defects, including neurological outcomes, joint abnormalities, and face characteristics.

—Teratology Society (June 2016)

Fetal abnormalities caused by the mosquito-borne Zika virus first came to public attention in January 2016, the same month that many of us learned of a "29-inch-tall phocomelic overachiever" named Matthias Buchinger (1674–1739). For this we can thank Ricky Jay, a popular magician of our day who found his avatar in decades of "Peregrinations in Search of the 'Little Man of Nuremberg.'" Ricky Jay paid a fitting tribute to his hero in an exhibit of his collection at the Metropolitan Museum of Art in New York, which accompanied a richly illustrated, companion volume. Jay has assembled a cabinet of graphic curiosities by and about this phocomelic overachiever, replete with portraits, broadsides, family trees, and coats of arms. Several advertise the Buchinger feats of skill at magic, musketry, skittles, and musical performance; most bear the telltale signature, "Matthias Buchinger born without hands or feet." Jay has the credentials to make Buchinger matter: Charles McGrath wrote in the *New York Times* that many consider Jay the "greatest sleight-of-hand artist alive, as well as a scholar, a historian, a collector of curiosities."

JAY HAS WRITTEN ABOUT BUCHINGER BEFORE, as one in a gallery of curious historical characters. In his 2016 volume Jay addresses Buchinger's mastery of micrography—the art of writing texts almost invisible to the naked eye—and explains how he accumulated a treasury of calligraphy by this tiny conjurer who had "two excrescences growing from the shoulder-blades, more resembling the fins of a fish than arms of a man." A connoisseur's passion convinced Jay that there was always more Buchinger material to be collected. His quest led him to dealers, print experts, and fans of micrography on both sides of the Atlantic. At one auction, Jay and Nicolas Barker of the British Museum competed for a grand collection but, because of its cost, were forced to divide the material. Jay tells us, "This proved surprisingly easy, as I took all the materials of armless calligraphers and Nicolas received the work of calligraphers conventionally digited."

Enlarged images in Jay's book show how Buchinger cunningly wove miniature texts into the crannies of calendars, portraits, and coats of arms. Two high-resolution images are exemplary. In Buchinger's posthumous portrait of Queen Anne (1718), limned in ink on vellum, the curls of the queen's hair are formed by micrographic letters that painstakingly spell out three chapters from the Book of Kings. In a stippled engraving from a 1724 self-portrait, Buchinger used micrographic lettering to form his florid wig: the text spells out seven complete psalms and the Lord's Prayer!

Jay tells us that viewers "respond to Buchinger's micrography as they do to the performance of magic: when they are stunned, or stumped, they seek an explanation." Biologists will not be surprised that Robert Hooke's name pops up here; Buchinger lived in the golden age of the microscope. Hooke's novel images of cellules in *Micrographia* (1675) and Antonj van Leeuwenhoek's pictures of "animalcules" in semen (1677) may well have prepared Buchinger's patrons for a new world of magic in magnification.

Jay was drawn to Buchinger not only for the little man's skill at micrographics but also by the story of a fellow conjurer who performed sleight-of-hand without hands: "magical tricks, performances on a variety of musical instruments, trick shots with pistols and swords, and bowling." From his childhood in Bavaria, where he was called a "thimble," Buchinger became increasingly famous over Western Europe as a skilled performer. He was a whiz at card play, swordplay, and even at dancing the hornpipe in a Scottish kilt. He entertained audiences at street fairs and manor houses, public inns and royal seats; he appeared in venues ranging from Leipzig to Dublin, the Tuileries in Paris to the Court of Saint James's in London. A 1726 broadside described him as "The Greatest German Living"—this in the reign of the Hanoverian George I.

Equally surprising was Buchinger's private life. Jay has quipped that this dwarf without arms and feet had at least one operating appendage: he managed to wed four wives and claim fatherhood of fourteen children. Buchinger's somewhat Freudian drawing of his family tree is neatly reproduced in Jay's book. Dated 1734, the two-page cutout depicts the artist's children—eight living, six dead; born in Zurich, London, or Dublin—as fruit hanging from the branches of a tall, thick paternal trunk. The tree is solidly planted on rootless lumps of soil atop the steles of four women, three dead and one living. Tiny flowers sprinkle the field, micrographic gametes in a pictorial autobiography. That "Greatest German" broadside described it thus:

> Great Trunk of Man be not ashamed
> That Nature has thy Body maim'd
> The Oak could not the Trophy bear
> Till that the branches cropped were,
> Nor would thy fame hae been so great,
> Had Nature formed thee quite compleat.

As for that "Greatest German" claim of fourteen seminal trophies, his contemporary Johann Sebastian Bach (1785–1850) had Buchinger beat. Formed "quite compleat" and conventionally digited, Bach fathered *twenty* children with two wives and had time left over for the *Mass in B Minor*. But Buchinger over his lifetime entertained audiences far more diverse in rank and geography than did the composer—and with neater penmanship to boot.

Readers figuring out how this "Body maim'd" became a paterfamilias will also wonder how Buchinger mastered miniscule calligraphy without the aid of optical gadgets. Ricky Jay was also puzzled by this and posed the question to several eminent artists. Their verdict came out on the side of lenses: Art Spiegelman, Erich Fischl, David Hockney, and Ed Ruscha guessed that Buchinger used magnifying lenses, possibly fixed to a ring-stand apparatus not uncommon at the time. Jay, writing "as a magician" concludes that Buchinger probably did ordinary lettering in public—a stunt in itself without digits—but worked out micrography in private, with magnifying lenses.

"Sure he did," Jay concludes skeptically, while remaining a fan of a fellow magician and author whose fame far exceeded his phocomelic bodily form.

RICKY JAY PAYS MORE THAN PASSING ATTENTION to that bodily form, another reason why biologists will want to read the book. Evolution dictates that in phylogeny, fins come before limbs: in phocomelia, limb ontogeny stops cold. Jay quotes eighteenth- and nineteenth-century authorities who supposed that phocomelia results from "maternal imprinting," then defined as "a traumatic stimulus encountered by any pregnant woman." Local officials even banned Buchinger from performing at fairgrounds, worried that women frightened by his appearance might bear children similarly malformed. It turns out that the idea of maternal imprinting isn't all that far-fetched. Zika, rubella, and other viruses linger in the tissues of the fetus and cause permanent, if not fatal, deformities. But I'd bet it wasn't Zika but a teratogen that did Buchinger in.

Epigenetic effects of thalidomide and vitamin A on our inner fish serve as a model of what happened to the little man from Nuremberg. Sainte-Hilaire's phocomelia was an anomaly for over a century after its description, but in 1957 along came thalidomide. Soon after the drug was approved in Europe to treat morning sickness in pregnancy, reports began to appear in medical journals of babies born with flipper-like limbs; the tabloids went wild. The drug was finally banned for pregnant women worldwide in 1961, thanks mainly to the courageous Frances Kelsey of the Food and Drug Administration, who had blocked its approval in the United States. Unfortunately, by then more than ten thousand children, mostly in Germany and Britain, had been born with drug-induced defects of their extremities; those who reached adulthood look very much like twenty-first-century Buchingers (an image of the German filmmaker Niko von Glasow is on Wikipedia). Work on the molecular pharmacology of thalidomide by Neil Vargesson has implicated a metabolite of the drug as a possible culprit and permitted discovery of the way its partner in tissues, the protein cereblon, acts to blunt limb development.

Both before and after thalidomide, another culprit has been implicated in phocomelia—at least in the lab. Following the work of Honor B. Fell and her students at the Stangeways Research Laboratory in Cambridge, studies of how high doses of vitamin A affect embryonic limb development have moved to the molecular level, as I described in an early paper. Excess vitamin A affects "maternal imprinting," but not by frightening expectant mothers. Recent studies, such as Sheikh et al., show that high doses of vitamin A produce changes in stem cell differentiation, acting in part by epigenetic modifications of pathways that dictate how fins become limbs in the course of evolution.

How would hypervitaminosis A account for limbless Buchinger? Prompted by the happily digressive tone of Ricky Jay's book, I'll make a bold suggestion. I doubt that Buchinger's mother munched on too many carrots; I'd put my money on an overdose of *Gruenkohl*—green kale, cabbage, or kraut. Kale, a staple of the Bavarian diet, has the highest vitamin A content of any food available in Buchinger's time. Could a Bavarian mother have been imprinted by an overdose of kraut? That notion seems as improbable as four wives and fourteen children fathered by a "little man from

Nuremberg." A trendy German website, facetiously named *Krautkanal,* celebrates Buchinger's "improbable" exhibit in New York, linking kraut virally to the dwarf. There is, of course, another improbable link between viruses and sauerkraut, which requires lactic-acid bacteria for its fermentation. Bacterial strains that have survived phage infection are the fittest for the job: to no one's surprise, CRISPR keeps the sauerkraut sour.

4.

Ike on Orlando: "Every Gun Is a Theft"

> Every gun that is made . . . signifies, in the final sense, a theft from the sweat of its laborers, the genius of its scientists, the hopes of its children.
>
> —Dwight D. Eisenhower (1953)

> In the councils of government, we must guard against the acquisition of unwarranted influence, whether sought or unsought, by the military-industrial complex.
>
> —Dwight D. Eisenhower (1961)

> If there weren't so many of these [handguns] around, why, maybe you could be a little more peaceful.
>
> —Dwight D. Eisenhower (1958)

IN THE DECADE OF SANDY HOOK, San Bernardino, and Orlando, the warnings of our thirty-fourth president against gun violence have been all but forgotten. TV commercials have featured car ads urging us to "celebrate Dwight D. Eisenhower's interstate highway system," while a film, *The Monuments Men,* praised Eisenhower's role in saving art treasures from Nazi plunder. Not to be outdone, the Metropolitan Museum of Art, the National Gallery, and the Smithsonian Institution displayed documents and works of art honoring Eisenhower. The Metropolitan Museum made him a Fellow for Life, citing him as "soldier, diplomat and statesman, through whose wisdom and foresight irreplaceable art treasures were saved for future generations." With all that attendant publicity, *The Monuments Men* posted the third-highest box-office results of its opening weekend, trailing only *The Lego Movie* and *RoboCop*. "President Eisenhower might have liked *RoboCop*," the critic Mick LaSalle commented, "because its villain was the same as his—the military-industrial complex."

Not a word about Ike's warning about the "hopes of its children" in the Sandy Hooks and Orlandos to come: "Every gun is a theft!"

Ike targeted other villains as well, including Joe McCarthy and his gang of rabid anticommunists. Ike came to mind after a gun-toting director of the National Rifle Association accused President Obama of being a "communist-nurtured subhuman mongrel." Ike had cautioned my graduating class of 1950 at Columbia College, "Let's not be stupid enough to fall into that grave error. Let's not call anybody a Communist who may be just a little bit brighter than ourselves."

When Eisenhower was first elected, in 1952, Senator McCarthy was riding high. Biographer Jim Newton reported that two years later, after the Army–McCarthy hearings, which Ike quarterbacked, the president could quip, "It's no longer McCarthyism. It's McCarthywasm."

SIX MONTHS AFTER HIS INAUGURATION IN 1953, President Dwight D. Eisenhower kept his election pledge by ending the Korean War. He had argued against any land war in Asia: "This world in arms is not spending money alone," he explained. "The cost of one modern heavy bomber is a modern brick school in more than 30 cities. . . . It is two fine, fully equipped hospitals." Not only bricks and mortar were at stake: the Korean War had taken 54,246 American lives and cost 320 billion of today's dollars. Ike would have appreciated a recent report from Brown University's Eisenhower Research Project—a name chosen to honor Ike's warning against a burgeoning military-industrial complex in the United States. The project reported that, in the first decade after 9/11, our wars in

Afghanistan, Iraq, and Pakistan had killed at least 225,000 people overall, among them 6,000 U.S. military deaths. Those wars have cost the United States between $3.2 and $4 trillion, including medical care and disability for current and future war veterans.

But it's not only guns abroad that kill Americans and rob us of our treasure. Would that Ike were around today to warn us against the non–military-industrial complex! Deaths and injuries due to gun violence have been estimated to cost our economy $40 billion annually. Meanwhile, ammunition manufacturers made a projected $993 million in profits on sales of $11.7 billion in 2013. That's over $400 billion for the decade on sales of $10 trillion, at a cost of 320,000 lives per year. Every gun is a theft: there have been more deaths from gun violence in the United States since 9/11 than all civilian and military deaths in Iraq, Afghanistan, and Iran combined. This level of violence, unique to the United States, constitutes a major public health hazard.

In 1955, Ike honored Jonas Salk for sparing countless thousands of American parents from the "agonizing fears of the annual epidemic of poliomyelitis." Kids were mainly at risk. There had been 38,741 cases of polio with 1,620 deaths in the previous year and an estimated death rate of 1 per 100,000. The White House established an advisory committee that was given broad executive power to assure fair, nationwide distribution of vetted lots of the scarce vaccine.

Well, these days we have an annual epidemic of gun violence, with a death rate ten times that of polio in 1954 and with twenty-fold more people killed each year. But in terms of public health, we are still at the pre-Salk level when it comes to gunshot violence. We know what causes the disease (bullets), we know the vector (guns, rifles, assault weapons, etc.), and we know where in the body the injuries lodge (watch any episode of *NCIS*). We actually know more about this condition than we know about threats such as cancer of the prostate or leukemia.

We also know some of the social factors involved. In the United States, kids are at risk: think Columbine and Sandy Hook. The death rate of blacks in the Unites States from firearms violence is twice that of the general population, with black youngsters ages15–24 particularly susceptible. Britain has a death rate from gun violence of 0.23; in the Bahamas, it's 22.2. If one compares two cities, Seattle and Vancouver, which have similar populations and overall crime rates, the Washingtonians were 4.8 times more likely to be killed with a handgun than their neighbors to the north. The Canadians have strict handgun control laws. We have a few clues to what might blunt the epidemic here in the United States: it's clear that the more gun dealers there are in any city, the more firearm deaths are recorded. It's also well documented that the more rigorous firearm laws are in any given state, the lower the rate of firearm fatalities. Conclusion: stay away from places with lots of guns, lots of gun dealers, and weak gun control laws. Avoid gun-toting Rocker/NRA bullies who still call President Obama a mongrel. But that advice is on a par with 1954 public health warnings to avoid *E. coli*–infested swimming pools during a polio epidemic.

Even before his naming of the military-industrial complex, Ike had a solid history of spotting bullies in the act of plunder. On May 26, 1944, eleven days before D-day, General of the Armies Dwight D. Eisenhower issued an unusual proclamation, quoted in full in Robert Edsel's *The Monuments Men.* Hoping to avoid the destruction of cultural heirlooms in the battle for Hitler's Fortress Europe, Ike gave his troops a clear command: "Inevitably, in the path of our advance will be found historical monuments and cultural centers which symbolize to the world all that we are fighting to preserve. It is the responsibility of every commander to protect and respect these symbols whenever possible." Eisenhower's command was widely respected; it also served as the marching order for the Monuments Men. Ike and his crew of GI monument wardens knew that they were dealing with art-hoarding Reichsmarschall Hermann Göring and that amateur dauber, Reichskanzler Adolf Hitler. In consequence, as Allied troops fought their way from the Channel beaches to the doors of Berlin, the Monuments Men paid attention not only to monuments but also to portable treasures filched from occupied countries and Germany's own citizens. Thanks to that D-day order, and to the devotion and service of the Monuments Men, thousands of treasures were saved from destruction.

On April 2, 1946, less than a year after war ended in Europe, General Eisenhower was honored at a gala reception hosted by the Metropolitan Museum of Art in New York. One of the Met's own Monuments Men, Director James Rorimer (played by Matt Damon in the movie), greeted his commander as Ike was hailed for his "oversight of the repatriation of artworks stolen by the Nazis during World War II." Eleven of these are now in the collection of the Metropolitan, given by families whose treasures had been saved; a tour of the lot is online today. Ike told the Met audience what the Monuments Men had unearthed the year before:

> There, in caves, in mines, and in isolated mountain hideouts, we found that Hitler and his gang, with unerring instinct for enriching themselves, had stored art treasures filched from their rightful owners throughout conquered Europe. Alongside bar and minted gold were found paintings, statues, tapestries, jewelry, and all else that the Nazis knew mankind would pay much to rescue and to preserve.

The moment is captured in a much-reproduced photo from the National Archives, which shows General Eisenhower at a salt mine near Merkers in the company of Generals Omar Bradley and "Blood and Guts" George Patton. At the Met he was in different company. Ike would serve the museum as trustee (1948–1953) and honorary trustee (1953–1969). His fellow trustees included Thomas B. Watson of IBM and Arthur Hayes Sulzberger of the *New York Times,* both also trustees of Columbia University. They saw in Eisenhower a future president of their university, and perhaps of the United States.

IKE, WHO HAD BEGUN TO DALLY with amateur oil painting, made reference in his museum talk to a fellow army artist, Bill Mauldin, the stellar cartoonist of *Stars and Stripes*. He said, "Frequently the soldier was led to express in artistic fashion something of his own reactions to the phenomena of war. At least you are acquainted with the efforts of our friend Mr. Mauldin in this regard, who spared no pains to show what he thought of us brass hats." Mauldin had ridiculed the spit-and-polish regulations General Patton imposed on his battle-weary troops. In return, old "Blood and Guts" had Mauldin hauled into his office for "sabotaging military discipline" and threatened "to throw his ass in jail." Mauldin's biographer, Todd de Pastino, tells the story. As soon as he heard the news, Eisenhower overruled Patton's censure, issuing a directive that no commanders were to interfere with the *Stars and Stripes*—including Mauldin's cartoons. Ike's stand on a free press was welcomed by the GIs, hailed by the press, and spread to a wider public in the postwar years. It confirmed Ike's reputation as a brass hat in touch with civil society: some saw it as a first step to higher office than military command.

Susan Eisenhower, Ike's granddaughter, tells us that, overseas in 1942, Ike wrote in his diary a short eulogy on his father, who had died back home in Abilene, Kansas: "His finest monument is his reputation." He could have been describing himself. When Ike returned to the United States, his reputation had soared as a leader of men who swept the Nazis from Europe in the world's greatest battle to date. He'd juggled the demands of prima donna generals such as Montgomery and Patton and thorny allies like de Gaulle; he had established interim civil regimes in the liberated countries and in partitioned Germany; he had exposed and opened the concentration camps; he'd attended to monuments and returned stolen treasures. As shown in his speech at the Metropolitan Museum, he was also a true humanist who spoke and wrote well: "They who have dwelt with death will be among the most ardent worshipers of life and beauty and of the peace in which these can thrive."

With a legion of fans behind him and active support by Republican stalwarts such as Watson and Sulzberger and progressive Democrats like James Roosevelt, the "Draft Ike" movement began as early as 1948 and hit its stride in 1952, as David Pietrusza points out. After a brief stint as president of Columbia University—to establish his bona fides as a civilian leader—Republican Dwight D. Eisenhower swept the Electoral College, 442 to 89, to become our thirty-fourth president.

Only the Dixiecrat states refused Ike their votes, but Ike made his position clear. In 1953, in his first State of the Union message, he declared: "I propose to use whatever authority exists in the office of the President to end segregation in the District of Columbia, including the federal government, and any segregation in the Armed Forces." He used that authority, and followed it up with the first civil rights legislation since Reconstruction. And then, of course, Ike presided over the interstate highways, the end of the Korean War, the start-up of NASA, the establishment of the National Medal of Science, and the pushback against "so many handguns out there." Finally, after two terms in Washington, Ike proudly stated: "The United States never lost a soldier or a foot of ground in my administration. We kept the peace. People ask how it happened—by God, it didn't just happen, I'll tell you that."

5.

Nobel on Columbus Avenue

The summer of 1997 was a busy one for phone lines, email connections, and delivery services between Baltimore and Worcester, with numerous collaborative experiments.

—Andrew Z. Fire, Nobel Lecture (2006)

News of those "numerous collaborative experiments" went viral after December 10, 2006, when King Carl XVI Gustaf of Sweden honored Andrew Fire of Stanford and Craig Mello of the University of Massachusetts with the Nobel Prize in Physiology or Medicine. The two scientists had identified RNA interference (RNAi), a defense mechanism used by many organisms against double-stranded RNA (dsRNA) viruses. Interfering RNAs have provided us with a means of selectively shutting down one or another genes in a cell and given us clues to antiviral therapies to come.

A few months later, the names of Fire and Mello joined those of other American laureates carved into a pink granite monument at the corner of Columbus Avenue and 81st Street in New York. The monument, dedicated to Alfred Nobel and to American Nobel laureates past and present, stands behind the American Museum of Natural History in a park named after Theodore Roosevelt, the first American laureate (Peace Prize, 1906). The names of 290 other American Nobel prize–winners have been inscribed on the sides of the monument, and there is ample space for more to come.

The monument in New York came to mind as I read Craig Mello's conversation with Adam Smith of the Nobel Foundation, who called Mello right after a call from Stockholm had given Mello the news of his award on October 2, 2006.

[ADAM SMITH]: Well first of all many, many congratulations on being awarded the prize.
[Craig Mello]: Thank you so much.
[AS]: Where were you when you heard the news?
[CM]: I was checking my daughter's blood sugar. She has type 1 diabetes so I was actually up, one of the few, I guess, in the North Americas who was awake.
[AS]: Yes, I imagine so.
[CM]: We check her frequently and I just happened to be up, checking her blood sugar. And she had a good sugar actually, 95, which is normal.
[AS]: That's good news, yes.
[CM]: I was on my way back to bed and the phone rang.
[AS]: So two good pieces of news at once! I imagine you were thinking of other things but what was your first thought on being told?
[CM]: Well, you know, gee, that's a really hard question! You know first it's disbelief, and I don't think it sinks in quickly. I felt I was sort of too young to get it this soon and thought, if it happened . . .

"This soon" and not later, 46-year-old Mello joined other striplings whose names were carved in granite, among them Joshua Lederberg and James Watson (ages 33 and 34, respectively, when they received their notifying telephone calls).

The Nobel monument was a recent addition to the New York scene. The slab was unveiled on a damp October day in 2003 by Mayor Michael Bloomberg, with appropriate remarks by Swedish and Norwegian dignitaries, choral music, and a stirring address by Eric Kandel (Physiology or Medicine, 2000), who paid tribute to the city's public school system, of which he was a product. In the audience were many Nobel prize–winners, their guests, consular and civic officials, and platoons of students from high schools in the neighborhood.

As the flock of VIPs dispersed, a teenaged couple made its way to the monument. Arm draped around his girlfriend's shoulder, a gangly youth pointed up at the open space under all those names: "I'm going to be the first black guy up there in science!"

That's why we have prizes and monuments, I thought at the time. To carve something in stone for a kid to look up to.

THE STORY OF dsRNA-MEDIATED gene silencing wasn't always carved in stone for a kid to admire. The work of Mello and Fire bears the hallmarks of a "seminal discovery," a term usually applied in retrospect and of three aspects. First, the unexpected often comes from a field far removed from the scientist's own field of expertise. Next, it stares one in the face, unexpectedly, like Mendel's sweet peas or Darwin's finches. Finally, it tips its hand when the seminal publication uses key words like "unexpected" or "surprise" in title or abstract. Here's how siRNA (small interfering RNA) was announced in the 1998 paper that Fire, Mello, and others published in *Nature*: "To our surprise, we found that double-stranded RNA was substantially more effective at producing interference than was either strand individually."

While the role of siRNA-mediated gene silencing is now written in stone, the path to its discovery was by no means straightforward: it began with petunias, led to worms, then finally on to yeast and beast. Andrew Fire recalled in his conversation with Adam Smith, "Well, we were led to it pretty much by our experimental noses. The people in the plant field had done tremendous work on gene silencing and so we, sort of, were following in their footsteps in trying to sort out what was responsible for this weird silencing phenomenon in the worm."

Their experimental noses had a long trail to sniff. Deciphering what George Sen and Helen Blau called "the silence of the genes" began with decoding the color purple in petunias and has wound up producing compounds of promise in disease. In 1990, plant scientists Carolyn Napoli,

Christine Lemieux, and Richard Jorgensen were studying the formation of anthocyanin, the pigment that makes petunias purple. Testing whether the enzyme chalcone synthase was the protein critical for the color purple, they overexpressed chalcone synthase (CHS) in petunias. In their paper they wrote, "Unexpectedly the introduced gene created a block in anthocyanin biosynthesis," and 42% of the plants became white or had chimeric purple–white patterns. They had silenced the gene for CHS, and this silencing proved to be heritable: "progeny testing showed that the novel phenotype co-segregated with the introduced CHS gene." Purple no longer, by heredity.

Sure enough, "unexpectedly" was the operative word in the petunia paper, and the work it described led directly to Fire and Mello's "surprise" at finding double-stranded RNA interference in worms. Many viruses contain double-stranded RNA, which, when injected into a cell, binds to a complex series of proteins (one called Dicer) that degrade viral RNA, and so the cell survives the infection.

These days, when I pass by that monument on Columbus Avenue, I'm reminded of petunias, and all those messages sent by "phone lines, email connections, and delivery services between Baltimore and Worcester." Going viral, indeed.

Planning for the Nobel monument in New York began with discussions in December 2001, when the Nobel Foundation celebrated its centenary in Stockholm. Most of the living Nobel laureates in Physiology or Medicine (including a good number of Americans: Joshua Lederberg, Barry Blumberg, Alfred Gilman, Joseph Goldstein, Michael Brown, Eric Kandel, among others) showed up to revisit their grandest moment. The occasion was celebrated by lectures, symposia, concerts and the grandest of all banquets in Stockholm's splendid Town Hall. My wife and I were guests, and we were lucky enough to have been seated at the banquet table with the senior American laureate present, Tom Weller of Harvard (1915–2008). He told a story that described one way to make a great discovery: make a great error, and correct it.

Over wines of not inconsiderable vintage, Weller reminded his dinner companions that the first time American scientists had been feted in this hall, it was for a dazzling cure based on luck and error. In 1934, two Harvard clinicians, George R. Minot and William Murphy, joined George C. Whipple, a Rochester pathologist, on the podium in Stockholm. From the wrong animal model, they had found a cure for pernicious anemia.

George Richards Minot, a patrician Bostonian, had been a professor of medicine at Harvard since 1928. He also maintained an active private practice oriented to hematology. William Parry Murphy, on the other hand, was a Westerner of decidedly nonpatrician stock. After one year at the University of Oregon Medical School he won a scholarship to Harvard Medical School, graduated in 1922, and in the midst of his Boston residency Minot made Murphy an offer he could not refuse.

Minot, as the story goes in *Medicine at Harvard,* was accustomed to picking young physicians of Peter Bent Brigham Hospital as associates to run his office practice, which consisted in good part of patients with diseases of the blood—and with homes on Beacon Hill. As senior resident, Murphy was next in line for this plum job, but it was customary for the young men to earn their credentials by publishing one or more papers before they started. Murphy took to the journals and found a recent report by George Whipple and Frieda Robscheit-Robbins that dogs made anemic by repeated bleeding could be restored to health by being fed huge quantities of uncooked liver. If it worked in dogs, why not in humans?

The first patients to whom Murphy fed slightly cooked liver were patients with pernicious anemia. One of these patients was an old woman who had recently become cranky and obstreperous, but whom, after a mighty contest of wills, Murphy cajoled into taking her daily ration of half a pound of liver. Murphy described his everlasting "surprise" that not only did her red blood cells respond to a week or so of this cumbersome regimen, but also that she was relieved of her mental symptoms. She soon reverted to her former, agreeable self. Eventually, his lucky observation of her mental improvement persuaded Murphy that the active factor in liver must work not only in the marrow but also elsewhere in the body. Iron alone could not do the trick.

By May 1926, Minot and Murphy had treated each of forty-five patients with half a pound of liver a day. Soon Minot persuaded a young biochemist, E. J. Cohn, to make a liver extract, which would be rich in the anti–pernicious anemia factor. The extract briskly revved up red cell production by the marrow and cured the disease as readily as raw liver. Indeed, it remained the primary treatment for pernicious anemia until Alexander Todd isolated and named vitamin B_{12} in 1948. It was among the first miracle drugs to cure a hitherto fatal disease.

Ironically, Whipple's dogs had not suffered from pernicious anemia—they had iron-deficiency anemia caused by repeated bleedings, and their response was due to the iron present in massive doses of liver. Whipple's error and Minot's luck led directly to the first Nobel Prize in Physiology or Medicine awarded for work in the United States.

There's more to the Minot story, and it touches on Mello's conversation with Adam Smith of the Nobel Foundation. Like Mello's young daughter, Minot suffered from what we would now call type I diabetes: he was gravely ill when he made his great discovery. A strapping six-footer, Minot had developed a severe case of diabetes, and by 1922 his weight had dropped to 120 pounds. As luck would have it, Charles Best and Frederick Banting in Toronto, following the approach of John Macleod (Medicine and Physiology, 1923) had prepared the first useful batches of insulin. Minot was in the initial group of patients treated with the hormone by Eliott Joslin, the first American specialist in diabetes. One Nobel led to another: had insulin not been discovered, vitamin B_{12} might have been a distant dream. It's what Robert Hooke predicted in 1665 when the Royal Society began its assault on "not precisely Knowing": "By this means they find some reason to suspect that those [phenomena] confessed to be occult, are performed by the small machines of Nature."

THE NAMES OF FIRE AND MELLO are now up there with those of Minot and Murphy, carved in stone on that pink granite slab behind a New York museum. More likely than not, some other kid from the neighborhood will point up at those names and start to wonder about what it takes to discover the new.

6.

Lupus and the Course of Empire

United nations shall combine,
To distant climes their sound convey,
That Anna's actions are divine,
And this the most important day
The day that gave great Anna birth
Who fix'd a lasting peace on earth.

—G. F. Handel, "Ode for the Birthday of Queen Anne" (1713)

George the Third
Ought never to have occurred.
One can only wonder
At so grotesque a blunder.

—E. C. Bentley, "George III," (1929)

WERE IT NOT FOR LUPUS and the antiphospholipid antibody syndrome, the United States might now have a National Health Service. That's my conclusion from finding links between the most prominent victim of the syndrome—Queen Anne—and recent debates over Scotland's secession from England. Anne (1665–1714) was the last of the Stuarts, a dynasty that gave England not only the union with Scotland but also the Royal Society, Isaac Newton, and the copyright law. The "grotesque" Georges who succeeded her led the British to Bunker Hill and Yorktown.

In September 2014, Scots held a referendum on whether to dissolve a partnership that had lasted since Queen Anne presided over the Acts of Union (1707), which established the "united Kingdom of Great-Britain." As Simon Schama wrote in *A History of Britain,* Anne saw to it that the United Kingdom, which "began as a hostile merger would end in a full partnership . . . " Anxious voters were reassured that, after independence, Scots would retain their popular National Health Service. Secession lost out, but Scotland's first minister had already set Scottish Independence Day as March 24, 2016, to commemorate the anniversary of the Acts of Union and the United Kingdom's founding mother, Queen Anne. The queen, a childless widow, died in 1714, sickened by what was then diagnosed as gout, dropsy, hemorrhage, and stroke. If poor Anne had produced a Stuart heir, that blundering George from Germany would not have ascended to the throne, and the United States today might have a National Health Service like those in Scotland, Canada, or Australia.

QUEEN ANNE'S LIFE AND THE STUART DYNASTY were undone by systemic lupus erythematosus (SLE) and its harsh companion, the antiphospholipid antibody syndrome, which produces bleeding, clotting, stroke, and obstetrical calamity (Hughes, 1985). Anne and her husband, Prince George Oldenburg of Denmark, sweated out at least seventeen pregnancies from 1684 to 1700 (Table 1): all but one resulted in miscarriages, stillbirths, or infant death. Anne's longest-surviving child, William, the last Stuart to live at Kensington Palace, died at age 11 after suffering infantile seizures, childhood dyskinesias, and gross hydrocephalus, symptoms now recognized as those of neonatal lupus.

Table 1. Children of Anne Stuart, Queen of Great Britain, and George Oldenburg, Prince of Denmark

Children
Stillborn daughter 1 Oldenburg b. 12 May 1684, d. 12 May 1684
Mary Oldenburg b. 2 Jun 1685, d. 8 Feb 1687
Anne Sophia Oldenburg b. 12 May 1686, d. 2 Feb 1687
Stillborn child 1 Oldenburg b. 21 Jan 1687, d. 21 Jan 1687
Stillborn son 1 Oldenburg b. 22 Oct 1687, d. 22 Oct 1687
Stillborn child 2 Oldenburg b. ca. Oct 1688, d. ca. Oct 1688
William Henry Oldenburg, Duke of Gloucester, b. 24 Jul 1689, d. 30 Jul 1700
Mary Oldenburg b. 14 Oct 1690, d. 14 Oct 1690
George Oldenburg b. 17 Apr 1692, d. 17 Apr 1692
Stillborn daughter 2 Oldenburg b. 23 Mar 1693, d. 23 Mar 1693
Stillborn daughter 3 Oldenburg b. 21 Jan 1694, d. 21 Jan 1694
Stillborn daughter 4 Oldenburg b. 17 Feb 1695, d. 17 Feb 1695
Stillborn son 2 Oldenburg b. 25 Mar 1696, d. 25 Mar 1696
Stillborn son 3 Oldenburg b. 25 Mar 1697, d. 25 Mar 1697
Stillborn son 4 Oldenburg b. 10 Dec 1697, d. 10 Dec 1697
Stillborn son 5 Oldenburg b. 15 Sep 1698, d. 15 Sep 1698
Stillborn son 6 Oldenburg b. 25 Jan 1700, d. 25 Jan 1700

It's clear that Anne suffered from SLE, a disease that chiefly afflicts women of child-bearing age and their newborns. The queen's contemporaries describe four clinical features that add up to current criteria for the diagnosis, described in Petri et al.: a blotchy, pitted face with a wolf-like rash on the cheeks; recurrent polyarthritis; facial and leg edema; and repeated seizures, nosebleeds, and lethal stroke. Official portraits of Queen Anne show variable joint swellings, obvious facial swelling, and the classic facial rash. Add her obstetrical history, and we arrive at the diagnosis of the antiphospholipid antibody syndrome. The syndrome is often tagged "Hughes syndrome," after my colleague Graham R. V. Hughes, who described a patient in London with ailments similar to those of Queen Anne. His seminal 1983 article in the *British Medical Journal* sums up the problem: "Thrombosis, abortion, cerebral disease and the lupus anticoagulant." Graham Hughes has earned his eponym: it's fitting that he directs a unit at St. Thomas' Hospital, down Royal Street and a bridge away from Westminster Abbey, where Queen Anne lies forever.

Hughes syndrome, which can also occur in the absence of lupus, results from antibodies directed against one's own cell membranes (phospholipids and/or proteins). Closely related antibodies, the "lupus anticoagulant," inhibit the coagulation of normal blood; in pregnant women, these autoantibodies induce cascades of injury directed against the mother or products of her womb. A clue to the diagnosis, as in the queen's case, is a history of fetal loss.

Before the reign of Queen Anne (1702–1714), England was torn by religious and family spats that ranged in intensity somewhere between today's Sunni–Shiite conflicts and intramural White House squabbles. The House of Stuart regained power in 1660 after Cromwell's Puritan misadventures. Anne's uncle, Charles II, restored the monarchy, presided over Restoration comedy, and chartered the Royal Society. The second Charles was a middling Protestant, as Anne had been brought up. But next in line to Charles came a very partisan Catholic, his brother, to be James II, who was Anne's father. After a short three years in power, James was overthrown in 1683 by Anne's Protestant brother-in-law (and cousin), who became William III. This coup, called the Glorious Revolution of 1688, resulted in permanent exclusion of any Catholic successor. William III became joint monarch with Anne's elder sister, Mary II, whom he had married—we know the pair from the college in Williamsburg, Virginia, whose name honors them. As a website devoted to British royals puts it, "the marriage survived although all three of her pregnancies were stillborn." Two sisters, sixteen stillbirths, no heirs? Time for some genomics here.

Then came the younger sister's turn: Queen Anne with her own stillbirths, her gout, dropsy, and seizures. But these days her reign is remembered less for disease than for peace and prosperity. The War of the Spanish Succession had broken out on both sides of the Atlantic the year before her coronation. The great powers—England, Austria, and Holland versus France and Spain—fought battles from the Alps to the Canaries, from

Jamaica to the Arctic. Handel's musical tribute to the "great Anna" celebrated her major achievement, the Treaty of Utrecht (1713). The treaty not only established a peace that would last to midcentury but also left Britain in possession of Newfoundland, Nova Scotia, the Hudson Bay Territory, and Gibraltar. Schama had it right—that full partnership of the United Kingdom had become "the most powerful going concern in the world."

Anne followed a path set by uncle Charles II as custodian of arts, science, and the commonweal. She was a patron of Christopher Wren, knighted Isaac Newton in Cambridge, and appointed Jonathan Swift the dean of St. Patrick's in Dublin. By proclaiming the "Statute of Anne" (1710) for the "Encouragement of Learned Men to Compose and Write useful Books," she established the basis of copyright law in anglophone countries. In the American colonies, her contented subjects commemorated her name and the benefits of her deeds. Annapolis, Maryland, is named for her, as are Cape Ann in Massachusetts and Fort Ann in Washington County, New York. She is remembered for granting an Act of Denization (by which a foreigner obtained legal status) to Luis Gomez, a Jewish refugee from the Spanish Inquisition in 1705. This document allowed Gomez to conduct business, own property, and live freely within the colonies. His mill in Marlboro, New York, is a tourist site today. Among her other acts, deeds, and grants that remain in the news are those 215 acres the queen bestowed on Trinity Church in Manhattan in 1705. The church elders are debating what to do with the $2 billion it's worth today.

Not bad for a dozen years of Stuart-ship, and again one wonders what a living heir would have meant.

Trouble came when the Hessians followed the Stuarts. Worried over Anne's afflicted womb, Parliament passed the Act of Settlement (1701), which assured a Protestant line of succession. The nearest skein of that line led to Hanau (Hanover) and the four Georges who ruled from 1714 to 1830. George I, a Hessian who barely spoke English, kept several mistresses; in return, his wife eloped with a Swedish count, who was killed and dumped in a river on George's order. He then had his young son, George II, arrested for siding with his mother and excluded him from public ceremonies. When his father died of a stroke on one of his frequent trips home to Hanover, George II assumed the British throne and—one generation after Anne's "lasting peace"—took the country to war again. The issue was settled in 1745, when Bonnie Prince Charlie and his Highlanders were defeated by the redcoats at Culloden. In 1751, George II's eldest son, Frederick, died suddenly of mysterious injuries (having been struck by a tennis, or possibly a cricket, ball), and the crown passed to George III.

At age 22, George III became head of the British Empire in 1760. The official website of the British monarchy notes that he is best remembered for provoking American independence and for going mad—adding, "This is far from the whole truth." Alan Bennett's popular play and the film made from it, *The Madness of King George* (1994), revived the story of a nutty monarch crazed by "variegate porphyria." Modern analyses reject that diagnosis but not its symptoms: blindness, deafness, and madness—episodic bouts of which followed the loss of the American colonies. When his redcoats and Hessian mercenaries were defeated by the Americans, he declared a General Day of Fast in 1778—a gesture understood as pitiful at the time. Horace Walpole (1717–1797) wrote:

> First General Gage commenc'd the war in vain;
> Next General Howe continued the campaign;
> Then General Burgoyne took the field and; and last
> Our *forlorn hope* depends on *General Fast.*

Whether his madness was caused by, or was coincident with, loss of his American colonies remains in doubt. What is certain is that George III blundered into his American quagmire through economic miscalculation. The empire was going broke, thanks to the costs of the successional wars with France and Spain and the expenses of the East India Company, which ran India for the crown. By the 1770s, at a time when there were no income taxes, the United Kingdom required £4 million (£500 million today) simply to service its debt. The answer was to tax items in demand among the more prosperous of the American colonies. George had figured out a solution. In a letter of the early 1780s, he wrote, "While the Sugar Colonies [the

Caribbean] added above three millions a year to the wealth of Britain, the Rice Colonies [South Carolina, etc.] near a million, and the Tobacco ones [Maryland, etc.] almost as much; those more to the north [Pennsylvania on up], so far from adding anything to our wealth as Colonies . . . rivalled us in many branches of our industry, and had actually deprived us of no inconsiderable share of the wealth we reaped by means of the others."

The answer was clear: impose taxes on sugar, tea, and commercial transaction. The British were sure that those moneymaking rivals would return some of the "not inconsiderable wealth" in the form of taxes. The result of that miscalculation was the American Declaration in Philadelphia of July 4, 1776, which lists in detail a litany of " . . . the patient sufferance of these Colonies" and explains the "necessity which constrains them to alter their former Systems of Government."

We do not know whether a legitimate heir of Queen Anne would have forestalled rebellion in Scotland or revolution in America, but I can imagine a placid Anne or regal Charles on the throne, making sure of "a lasting peace" on both sides of the Atlantic. Without those antiphospholipid antibodies tugging at Anne's womb, the Georges might have remained in Hesse, and the United States would have a National Health Service, just like Scotland.

7.

Groucho on the Gridiron

> After taking a sociology exam, Cardale Jones, a quarterback at Ohio State, posted a message on Twitter . . . : "Why should we have to go to class if we came here to play FOOTBALL, we ain't come to play SCHOOL, classes are POINTLESS."
>
> —*New York Times*, December 30, 2014

> Robert Morris University becomes first to recognize video games as varsity sport . . . scholarships will [be] worth up to $19,000 per student.
>
> —Associated Press, October 6, 2014

> ZEPPO: . . . it isn't right for a college to buy football players!
> GROUCHO: (president of Huxley College): This college is a failure. The trouble is, we're neglecting football for education.
>
> —*Horse Feathers* (film), 1932

THE STORY OF "ATHLETICS" IN AMERICAN COLLEGES has never been more amusing. Shortly after football players (undergraduates!) at Northwestern University demanded a union contract, Robert Morris University advertised that it would award "athletic scholarships" [*sic*] for varsity video-game players. Antics like these would have been enough to startle folks at universities in Stockholm, Edinburgh, Cambridge, or at the Sorbonne—schools that fulfill their academic functions without athletic scholarships or stipends for electronic gaming. But, not to be outdone, the National Collegiate Athletic Association (NCAA) raised the ante by offering a few thousand unaudited expense dollars to each American college athlete's already generous football scholarship. The NCAA proceeded to slip in a payment to ESPN for $7.3 billion over twelve years. It guaranteed telecast of seven "collegiate" games per year: four major bowl games, two semifinal bowl games, and the "College Football Playoff National Championship Game."

That contest—the first of its kind—featured quarterback Cardale Jones of Ohio State University, the fellow who said he "ain't come to play SCHOOL." Cardale came "to play FOOTBALL," and played the game of his life as his team beat Oregon 42–20. Jones showed the world that he'd learned his subject well and will be far better off than if he'd taken a shine to academics. Brawn pays the recent graduate a lot better than Brain: while the average English major expects about $32,000 a year on graduation, Jones is on track to be a millionaire in the NFL the minute he says "Goodbye, Columbus."

Most folks I know are "academics," a name that harks back to a lush grove of olive trees: the Akademia at the northern edge of Athens. Dedicated to the goddess Athena, the grove was home to the school of philosophy founded by Plato around 380 B.C.E. The Socratic dialogues and symposia held in the Grove of Academe have remained a model for scholarly discourse to the present day. There, too, Plato first defined the liberal arts. Horace looked back in wonder at that Athenian grove where scholars might "*inter silvas academi quaerere verum*" ("seek for truth among the trees of Academe"). Nowadays we honor the Grove of Academe as the birthplace of reasoned inquiry, the font of Brain.

Brawn, however, was honored there as well. The grounds of the Akademia harbored a gymnasium and bathing area, by the side of which Plato held forth from an exedra, a pillared courtyard, where wine and bathwater flowed for gymnast and philosopher alike. However, even before Plato, the Grove of Academe was renowned as the starting point for festivities that preceded the quadrennial Panathenaic games. Under torches lit from the altar of Eros, lively processions moved into town by the Dipylon Gate to glorify the contests: torch races, sprints, chariot races, wrestling bouts, javelin flings, and more. Competition was fierce, and athletes were well rewarded. Athletes placing first or second in each category received large Panathenaic amphoras, or jugs, that depicted the event; these overflowed with 40 liters of first class olive oil ($185 per liter today). The oil was meant to be sold and could be exported free of duty; the number of amphoras given as a prize depended on the event, the age category, and the final standings. A young (collegiate-level) athlete who placed second would get only six amphoras, but an older pro (NFL-level) who came in first would be given sixty of the jugs. Amphoras that survive today are priceless, the pride of major museums worldwide.

Plato was of one mind with Ohio's Cardale Jones. The philosopher held that intellectual and athletic efforts were equally arduous; he stated, "An athlete who aims for an Olympian or Pythian victory— he has no free time for anything else." But Brain and Brawn were not equally rewarded: athletes got the finest oil, Socrates got the deadliest poison.

WE CAN SEE THAT BRAIN AND BRAWN are similarly rewarded today, using Ohio State University and its Buckeyes as examples. Ohio State is the very model of a modern Big Ten university, ranked first in its state and eighteenth among American public universities overall. Its salary scale fits the nationwide model as well: NCAA Class 1 football coaches are paid approximately three to four times as much as university presidents.

Nevertheless, and to its credit, Ohio State's College of Art and Sciences—in which quarterback Jones is enrolled—in a "Statement of Principles" stresses the importance of a liberal arts education that "should not be compromised for the sake of expediency in pursuit of acquiring vocational skills." Accordingly, a list of Ohio State University people prominent in the arts includes Milton Caniff, James Thurber, and Roy

Lichtenstein, and the sciences have been enriched by three Nobel laureates with Buckeye roots: Paul Flory (Chemistry, 1974), Kenneth Wilson (Physics, 1982), and William Fowler (Physics, 1983).

Ohio State undergraduates can pick from a menu of 80 majors that range from Arts Management to Zoology; the longer list of 100(!) minors includes Coaching Education, Fashion and Retail Studies, Meat Science, and Turfgrass Management. So much for Plato's academy, which offered students the quadrivium: arithmetic, geometry, astronomy, and music. However, if a Buckeye can get a bachelor of science degree after studying Turfgrass Management (B.S. seems right), why not skip classes altogether and work on the turf itself? That would be a FOOTBALL major, and you ain't gotta go to class. It's also a modern version of Plato's recipe, "no free time for anything else," for winning an Olympian victory or playing in the College Football Playoff National Championship Game against the Oregon Ducks.

On January 12, 2015, the championship game was viewed by 80,000 fans in a Texas stadium and by over 30 million fans on TV. That's Darwinian survival of the football fittest over those who come to school to go to class. We note that the Massachusetts Institute of Technology football team played before 900 family and friends in the college stands, despite a record of 9–0 in 2014. The MIT quarterback was majoring in aerospace engineering.

Professionalization of American college football closely followed professionalization of the curriculum. For much of the nineteenth century, an Oxbridge-inspired program of classic education, plus or minus theology, was compulsory in English and American universities, with social Darwinism in the wings. In England, Thomas Henry Huxley, Darwin's great champion, argued in 1868 that for the sake of the empire, its gentlemanly college curriculum must be replaced by the sort of entrepreneurial program launched by the Germans: *Erdkunde*, " . . . a description of the earth, of its general structure, and of its great features—winds, tides, mountains, plains:

of the chief forms of the vegetable and animal worlds, of the varieties of man." Agriculture and soil management, as Ohio State would have it today.

The very next year, President Charles W. Eliot of Harvard University sniffed the spoor of social Darwinism and put Huxley's proposal into action. America was on the move after the Civil War, and expansion into the western territories called for practical education in the arts of empire. Pointing to European success at technical education, President Eliot proclaimed, "We are fighting a wilderness, physical and moral, and for this fight we must be trained and armed . . . " It may sound like social Darwinism these days, but it was a call for the elective system, which put an end to the classic core forever. Thanks to Huxley and Eliot, elite universities on both sides of the Atlantic replaced the study of Latin, Greek, and the quadrivium with elective courses in the natural and physical sciences, modern languages, sociology, geology, and engineering. To the sounds of that beat, students have been naturally selecting their own majors ever since.

Within a month of Eliot's call for manly vigor at Harvard, Princeton and Rutgers played the first intercollegiate football game in November 1869 before a rapt crowd of 200. That initial fight for survival of the collegiate fittest spread rapidly from coast to coast. Colleges accustomed to recruiting for academic excellence (and family standing) soon turned to recruiting for athletic skill (plus or minus family standing). The turning point came when the University of Chicago hired Alonzo Stagg as the first professional college football coach in the country—only two years after the university was founded in 1890.

Stagg, now honored as a patron saint of the coaching cult, founded an athletic dynasty at the University of Chicago. His brawny football team, the Maroons, earned its alliterative nickname "The Monsters of the Midway" by dominating its opponents in two out of three matches he coached. Entrepreneurial savvy came with football know-how. To boost sideline enthusiasm, the Maroons were first in the nation to wear varsity letters, and a mere two years after he was hired, Stagg gained national attention for the Maroons with a trip to Stanford, beginning the tradition of cross-country

"Bowls." Under Stagg, the University of Chicago was a founder of the Big Ten conference in 1895 and instrumental in putting together the NCAA in 1906. For half a century, Stagg's business acumen carried the day. The modest athletic stadium of his day, first known as Marshall Field ("Marshall" as in the department store), was soon renamed Stagg Field, eventually holding as many as 50,000 weekly fans of the Midway Monsters. The Chicago temple of Brawn came to an end when football was abolished in 1939 by Chancellor Robert Hutchens, a champion of the seven liberal arts.

A sidebar: The end of football at Stagg Field marked the beginning of the atomic age when, under the abandoned west stands, Enrico Fermi monitored the first nuclear chain reaction. *Sic transit,* as the Brains would have it.

Alonzo Stagg's legacy is huge. The billion-dollar football world of Cardale Jones and others today became what it is thanks to Stagg's invention of that uniquely American institution, the football scholarship. It was called a "student service payment" in the 1890s and evolved to guarantee survival of the strongest on the field, if not the classroom. Poor kids, rich kids, and kids of any color or origin, one and all, could go to college as long as they had Brawn—plus or minus Brain. Before "student service payments," tryouts for the team were open to any student who could handle a ball, as long as parents paid tuition. On to the era of Stagg, who was not always scrupulous in his selection of players. Hugo Bezdek, one of the 1904 Monsters of the Midway, was identified as a professional boxer named "Hugo Young" by a rival from the University of Illinois. Years later, Bezdek recalled to the *Detroit Free Press* that his Illini accusers had plotted his exposé at "some saloon," adding: "I don't know anything about the Illinois teams hiring the iceman to play for them . . . I don't know whether they really were a gashouse gang or not. They may have been genuine Illinois students."

So, college football players can do business at "some saloon"? Icemen and gashouse workers can be hired to play college ball? The prestige of a university can hang on victory over a rival at sport? It sounds like a recipe for a Hollywood comedy. It was.

That 1932 comedy is called *Horse Feathers* and forms a wicked footnote to a Carnegie Foundation report that had blown the whistle on a generation of collegiate shenanigans: "College football is a highly organized commercial enterprise. The athletes . . . are commanded by professional coaches. . . . The great matches are highly profitable enterprises," the report stated. In the flick, Groucho Marx plays Quincy Adams Wagstaff, the brand-new president of Huxley College. Groucho's son Frank, played by Zeppo, had been an undergraduate football player at Huxley for twelve(!) years. Groucho tells Zeppo that Huxley hasn't won a ball game for decades and that the college's reputation is on the line. The only way to save Huxley is to hire two hulking football players who hang out at a '32 speakeasy—"some local saloon," as in Stagg's Chicago. Alas, as luck would have it, the president of Darwin College, Huxley's archrival, had got to the saloon before him, and the two real hulks were gone. Groucho had to settle for two saloon regulars: Chico, a boot-legging iceman; and Harpo, an iceman and dogcatcher (shades of the Illini gashouse). Back on campus, university president Groucho took on the additional tasks of football coach, guidance counselor, locker-room trainer, and biology professor. Meanwhile, Chico and Harpo, like Cardale Jones today, went on to do what they were hired to do: to play FOOTBALL and to show that "classes are POINTLESS":

> Groucho (as a biology professor): Let us follow a corpuscle on its journey . . . Now then, baboons, what is a corpuscle?
> Chico: That's easy! First is a captain . . . then a lieutenant . . . then is a corpuscle!

The climax of the film, as expected, comes at the annual Thanksgiving football game, which pits Huxley against Darwin. All four Marxes are suited up, Groucho the president dons helmet, knickers, and cleats and goes on to make a tackle from the sidelines. It's a tight match, but Huxley has a selective advantage in this struggle for life. In the final quarter, bearing several concealed footballs, the four brothers are carried into the end zone in a chariot: Harpo's horse-drawn garbage wagon. Huxley wins 31–12: Brain beat Brawn in Academe, but it's only a flick. That image of four comedians riding across the goal line, *Ben-Hur*–style, was featured on the cover of *Time* magazine. It made the point that, in 1932, you could buy football players at a saloon and that the Carnegie Foundation was right: "The great matches are highly profitable enterprises."

THE FOOTBALL ENTERPRISE is perhaps even more profitable these days. There's that $7.3 billion contract signed between the NCAA and ESPN. As for the athletes, I'd argue that they should be rewarded for their Brawn and not be forced into the realm of Brain. Footballers in Division 1— as in the Big Ten—are on a professional career path and should be unionized; they should also be permitted to major in football itself. Like the Olympians who trained in the groves of Plato's Academe, the players have "no free time for anything else" and should not be forced to go to classes the principles of which they can acquire on the field. If you can major in Turfgrass Management, well . . .

Cardale Jones had it just right after the championship. He told the *New York Times*, "I don't think it's going to be based on your athletic ability. It's going to be based on your ability to process and diagnose information." Shades of Huxley—Thomas Huxley, the Darwinist, that is. I'd argue that a Division 1 football player should be free to pick any class he chooses that can teach him anything more valuable than "to process and diagnose information." As for the rest of us in Academe, how about the seven liberal arts for a start?

8.

Apply Directly to the Forehead: Holmes, Zola, and Hennapecia

> There is nothing men will not do, there is nothing they have not done, to recover their health and save their lives. They have submitted to be half-drowned in water, and half-choked with gases, to be buried up to their chins in earth, to be seared with hot irons like galley-slaves, to be crimped with knives, like cod-fish, to have needles thrust into their flesh, and bonfires kindled on their skin, to swallow all sorts of abominations, and to pay for all this, as if to be singed and scalded were costly privilege, as if blisters were a blessing, and leeches were a luxury.
>
> —Dr. Oliver Wendell Homes (1871)

Were Dr. Holmes to observe bodily mischief today, he'd still find needles thrust without cause into flesh and bonfires needlessly kindled on the skin. But nowadays the injuries are far less likely to be inflicted on the sick in search of health than on the vain in search of fashion. Botox bruises the foreheads of matrons, collagen scars the lips of barflies. Steel grommets hang from the navels of nymphets, bolts pierce the lips of perps. Perhaps the broadest practice, however, is the application of henna directly to hair and skin. This global assault has produced rock concerts that resemble the coming of age in Samoa and turned South Beach into the South Pacific. Warriors of the NFL sport body tattoos that put Papua to shame, while trendy folk in SoHo flaunt the umbilical baroque. If the Belle Époque was the Age of Gold, ours has become the Age of Tool and Dye.

Yet the medical literature documents that neither body piercing nor henna is all that safe. Injuries provoked by cosmetic intrusion spare no age, no gender, no color, no class. Even the very young fall victim, as reported in a 2007 news item headlined "Scarred Children":

> Michelle Lolk, of River Edge, took her 6-year-old daughter and 8-year-old son to a tattoo shop this past summer for their first-ever temporary tattoos. Young Ethan got a cross on his arm. His sister, Olivia, got a dolphin on her belly. A day later, Olivia complained of severe pain. "It looked like she was branded with a poker," Lolk said.

Skin branding of this sort (bonfires kindled on their skin, as Dr. Holmes might say) is due to acute contact dermatitis induced by henna's active agent, lawsone (2-hydroxy-1,4-naphthoquinone) and an added ingredient, PPD (para-phenylenediamine). Henna is a shrub (*Lawsonia inermis,* or Egyptian privet) cultivated in India, Sri Lanka, and much of North Africa. The dried leaves are mixed with various solvents and applied directly to the skin or hair. PPD is often added to red henna powder to produce the "black henna" preferred for tattoos. But PPD also renders the mixture more allergenic and sometimes virulently toxic: see the 1996 *Lancet* paper "A Woman Who Collapsed after Painting Her Soles." Temporary henna tattoos—the sort applied at rock concerts and kiddie festivals—are intended to persist for only a few weeks, but the incidence of acute inflammation, permanent scarring, and keloid formation has become epidemic in the last decade and a half. PubMed lists only three reports of reactions to henna tattoo in the two decades between 1975 and 1995—but 259 papers since 1995! A number of these cases were caused by henna without added PPD. Hair dyes come in all sorts of proprietary formulations: a 2006 study from Korea reported that of 15 henna samples tested, PPD was present in 3, nickel in 11, and cobalt in 4.

Henna has been recognized as an occupational hazard in hairdressing salons. At various doses the dye induces hemolytic anemia in lab animals and humans alike, and oral intake of henna produces an acute inflammation of the colon. In cell culture assays, lawsone causes cell death and cell cycle arrest. As might be expected for a redox-dependent naphthoquinine, individuals who lack oxidant defenses on a heritable basis, such as those with G-6 PD deficiency, are particularly at risk for Heinz-body hemolytic anemia.

In response to injuries caused by "temporary tattoos," the FDA last year issued a warning against the import of henna preparations containing PPD, but stated that the agency was powerless to supervise ingredients in "cosmetic samples and products used exclusively by professionals—for example, for application at a salon, or a booth at a fair or boardwalk."

So much for the kiddie trade!

WHILE THE YOUNG ARE APT TO APPLY HENNA directly to the skin, folks of a certain age mainly use henna to color their hair. The practice has been common for centuries in every corner of the world. Recently, however, France seems to have swept the honors for turning henna into art, as any stroller through Paris can attest. When the weather turns balmy, the streets are alive in a blaze of henna, offering coiffures in orange, auburn, red, and crimson.

But there may be a real downside to this display of vegetal finery. Over the years, I've observed on Parisian boulevards, in theaters, concert halls, cafés, and flea markets, a peculiar pattern of baldness in the French henna crowd. Women with hennaed hair, if of a certain age, seem almost uniformly to suffer from drastic, central alopecia—hair loss—quite evident at the back of the scalp, and quite noticeable in areas where the hair is parted. As a rheumatologist, I was struck by the difference between what one might call "hennapecia" and the commonly observed hair loss in patients with systemic lupus erythematosus. Nor is the pattern of hennapecia like that of ordinary female aging with its "increased thinning over the frontal/parietal scalp, greater density over the occipital scalp, retention of the frontal hairline, and the presence of miniaturized hairs."

Alas, the phenomenon is not limited to France. Although hennaed hair is less common in the United States, the same pattern seems to rear its ugly head, so to speak. In the fall of 2015, at the Pier Antiques Show in New York, where henna is also much in evidence, I observed forty-two women (approximate ages over 45) with hair overtly dyed with henna. Twenty-nine had clear signs of hennapecia. As a control, I observed thirty-six "blonde" women, blondness presumably due to peroxide, of the same age: only five had similar areas of hair loss. In both groups, the incidence of exposed, undyed roots was pretty much the same.

Ever since Lewis Thomas and I consistently produced hair loss in rabbits given excess doses of vitamin A, I have been intrigued by alopecia induced in lab animals and humans by agents of similar chemical structure. These instances have been associated with redox-induced changes in the hair growth cycle. Lawsone and PPD are clearly involved in redox cycling, and there are good reasons to believe that oxidative stress is involved both in graying and hair loss. This sequence was described by Arck and colleagues in the *FASEB Journal* in 2006. Their paper, "Towards a 'Free Radical Theory of Graying,'" concluded that "oxidative stress is high in hair follicle melanocytes and leads to their selective premature aging and apoptosis."

Whatever the cause of hennapecia, the condition cannot be due to acute inflammation or contact dermatitis: the many scalps I've observed, albeit at a distance, seem to have been uninflamed. It is, of course, entirely possible that women prone to one or another type of genetic hair loss have a unique recourse to henna, but my guess would be that the striking correlation between henna and hair loss puts the onus of alopecia on the dye and/or its additives.

It's clear that to settle the point, we need experiment, not simple observation.

The distinction between experiment and observation was spelled out for the general reader by Emile Zola (1840–1902) and brought to mind by a lurid postage stamp issued by the French Postal Service in 2003. Probably few of the eager philatelists who snapped the stamp up at first issue knew that the image was a poor caricature of a Manet portrait of Lucie Delabigne, the redheaded courtesan who became Zola's fictional Nana. Fewer still might have known that they were collecting a piece of scientific, as well as social, history. We can thank Claude Bernard, a founder of modern experimental medicine, for indirectly giving us Nana as well.

Emile Zola's 1880 novel *Nana* was a landmark of naturalist fiction, overtly based on methods spelled out in Claude Bernard's *Introduction à l'étude de la médicine expérimentale* (1865). In his *Le roman expérimental* (also published in 1880), Zola declared that

> Claude Bernard . . . explains the differences which exist between the sciences of observation and the sciences of experiment. He concludes, finally, that experiment is but provoked observation. All experimental reasoning is based on doubt, for the experimentalist should have no preconceived idea, in the face of nature, and should always retain his liberty of thought. . . . The essence of the higher organism is set in an internal and perfected environment [inherited characteristics] endowed with constant physico-chemical properties exactly like the external environment; hence there is an absolute determinism in the existing conditions of natural phenomena.

Zola assigned Nana a constant internal milieu, the "inherited characteristics" of a feral, manipulative courtesan. He then exposed his heroine to a variety of external stimuli (journalists, bankers, actors, gentry, tycoons), varied the strength of the buffer (theaters, garrets, hotels, mansions), and changed the ambient oxygen tension (age, war, disease, death). He recorded the results of these true-life interactions in the laboratory notebook of his naturalist novel.

Zola introduces Nana at time zero of his experiment. In her first appearance on stage, she is clad only in a diaphanous, see-through gown:

> Applause burst forth on all sides. In the twinkling of an eye she had turned on her heel and was going up the stage, presenting the nape of her neck to the spectators' gaze, a neck where the

> golden red hair showed like some animal's fell. Then the plaudits became frantic.

The experiment proceeds, her nature (feral, manipulative) remains constant; only the stimuli and buffers change, the lovers and venues vary. At the end of the book, after she has been exposed to a repertoire of environmental stress (age, disease, death), Nana dies of syphilis. A bellicose crowd marches under her windows at the Grand Hotel screaming "On to Berlin" as the Franco-Prussian War begins. Nana fades away, along with the Second Empire of Napoleon III: "On the bed lay stretched a gray mass, but only the ruddy chignon was distinguishable and a pale blotch which might be the face."

Zola described the faith of a nineteenth-century realist, in a passage that remains as pertinent to experimental biology as to the art of the novelist:

> The novelist is equally an observer and an experimentalist. The observer in him gives the facts as he has observed them . . . then the experimentalist appears and introduces an experiment, that is to say, sets his characters going in a certain story so as to show . . . the machinery of his intellectual and sensory manifestations, under the influences of heredity and environment, such as physiology shall give them to us.

Perhaps that's the machinery novels would still be exposing had novelists remained in touch with experimental science. And perhaps the FDA might send people to watch the influence of heredity and the environment on henna applied directly to the forehead.

Science Fictions

9.

Swift-Boating Darwin: Alternative and Complementary Science

I asserted, and I repeat, that a man has no reason to be ashamed of having an ape for a grandfather.

—Thomas Henry Huxley (1860)

AMERICAN SCIENTISTS BREATHED A COLLECTIVE SIGH of relief in December 2005 after Judge John Jones III ruled against teaching intelligent design (ID) in the classrooms of science. "The overwhelming evidence is that Intelligent Design is a religious view, a mere re-labeling of creationism and not a scientific theory," Jones declared in his 139-page decision, issued in Dover, Pennsylvania. "It is an extension of the Fundamentalists' view that one must either accept the literal interpretation of Genesis or else believe in the godless system of evolution. . . . The evidence presented in this case demonstrates that [intelligent design] is not supported by any peer-reviewed research, data or publications." But Dover isn't over.

Proponents of ID stubbornly refused to give up their campaign: "A thousand opinions by a court that a particular scientific theory is invalid will not make that scientific theory invalid," claimed Richard Thompson of the Thomas More Law Center, a group long devoted to Swift-boating Charles Darwin. The center had previously boasted that when ID had been inserted into the Dover science curriculum, "Biology students in this small town received perhaps the most balanced science education regarding Darwin's theory of evolution than any other public school student in the nation." Robert Crowther, director of the Discovery Institute, a Seattle-based think tank, so to speak, complained in a letter to the *New York Times* that Judge Jones's decision "asserts the factually false claim that ID proponents haven't published peer reviewed papers. A number of peer reviewed papers and books are listed on the Discovery Institute website at www.discovery.org/csc." William Dembski, a mathematician and Fellow of the Discovery Institute, insisted to the *Times,* "I think the big lesson is, let's go to work and really develop this theory and not try to win this in the court of public opinion . . . the burden is on us to produce." Demski, you've got a heck of a job to do.

The website of the Discovery Institute reveals that the "peer reviewed" evidence for ID consists of four articles. Each presents a theoretical argument that fails the test of experimental validation. Each has appeared in a publication devoted to pure speculation: the occasional *Proceedings of the Biological Society of Washington;* the thrice-yearly Italian/Indian review *Rivista di Biologia/Biology Forum;* the yearly *Dynamical Genetics;* or the every-other-year *Proceedings of the Second International Conference on Design & Nature*. We can conclude that active investigators of ID do not stoop to frequent or rapid publication. Nor does prestige dictate their choice of venue. Online HighWire Press, where seventy of the highest cited journals have been archived since 1948, lists Darwin's "natural selection" in 271 titles; almost all are experimental accounts. "Intelligent design" appears in the title of but a single effort, a conjectural review published in the *Journal of Theological Studies*. Alas, ID loses out to another system of alternative science: "Mesmer" or "Mesmerism" appears in 20 titles: each is devoid of experimental promise.

Although ID clearly lacks support in the literature of experimental biology, intelligent design remains a powerful notion that is no longer limited to extreme fundamentalists. ID may be coming soon to a science classroom in your neighborhood. At least ten states have legislation pending that would declare ID an alternative, or complementary, view to Darwinian evolution. And while Darwin's "theory" of evolution is as well accepted by scientists as the heliocentric theory of Galileo and the gravitational theories of Newton, it's easy to see why true believers resist the facts of common descent and natural selection. As Judge Jones decided, Darwinian evolution clearly contradicts "the literal interpretation of Genesis," and resolving that contradiction is difficult at best.

But I'm afraid that not only creationists or evangelists have questioned the experimental basis of science. The notion that there are alternative or complementary systems of medicine other than those based on the laws of physics and chemistry has swept not only daytime television, but also captured the hearts and minds of our legislatures and our elite universities and found a home on the campus of the National Institutes of Health. On an early web display, the National Center for Complementary and Alternative Medicine explained why it is funding work based on Ayurvedic notions of animal magnetism:

> This vital energy or life force is known under different names in different cultures, such as qi in traditional Chinese medicine (TCM), ki in the Japanese Kampo system, doshas in Ayurvedic medicine, and elsewhere as prana, etheric energy, fohat, orgone, odic force, mana, and homeopathic resonance.

While the Center is now called the National Center for Complementary and Integrative Medicine, its interest in vital energy and the "life-force" remains undiminished. Sadly, our current tolerance of homeopathic, chiropractic, Ayurvedic, holistic, crystal-based, or aroma-driven modes of healing has helped to clear the way for the alternative or complementary science of intelligent design. Once advocates of folk-based remedies persuaded the public that there are alternative or complementary explanations of what ails us, why not accept faith-based alternative or complementary explanations for how we came about? If the laws of chemistry and physics like the ideal gas law need not apply to medicine, why should we rely on the laws of evolution such as that of natural selection or the Hardy–Weinberg equation?

We live in an open, diverse society, disdainful since the 1960s of the hard facts of science. That disdain has both intellectual and religious origins: the intellectual roots are chiefly French, the religious roots American. On the one hand, the best and the brightest among us have been tutored in what Nicholas Kristof of the *New York Times* called the "Hubris of the Humanities." We have breathed the air of a postmodern era in which melancholic disciples of Michel Foucault proclaim "the end of our great faith in Progress." On the other hand, American science teachers in evangelical schools teach students that God created the world in six twenty-four-hour days.

No wonder that only 40 percent of Americans believe in evolution and that only 13 percent know what a molecule is. There are more who preach metaphysics than physics (49,230 clergy versus 16,680 physicists) in our country.

THE DOVER DECISION WAS A LANDMARK for those who value fact over faith in the realm of science; happily, there have been other such moments. Two public exhibitions (that happened to coincide with the Dover decision) reminded me of episodes as important to the life of science as that ruling by Judge Jones. The American Museum of Natural History mounted a comprehensive and compelling show on the life, work, and everyday impact of Charles Darwin. Illustrating Theodosius Dobzhansky's aphorism that nothing makes sense in biology except in the light of evolution, the exhibition attracted crowds of every age and hue in New York, and traveled to acclaim in Chicago, Toronto, and London.

Simultaneously, France celebrated 250 years of "Light and the Enlightenment" in a splendid exhibition at Nancy that served to illustrate Denis Diderot's argument that one can't traffic in metaphysics or morality without understanding the facts of natural science. It was organized by Jean-Pierre Changeux, the dazzling polymath of the Collège de France and our century's best friend of reductive science. The exhibit featured original texts, scientific artifacts, prints, and masterly paintings that documented the triumph of scientific light and reason over the forces of "*obscurantisme*" (the "Endarkenment"). Exhibits ranged from Newton's spectrum to Mme du Châtelet's equations to modern images of nerve conduction. It was good to see that in December 2005 the galleries at Nancy were as packed as the corridors in Manhattan or that courtroom in Dover.

The Darwin show in New York called attention to the famous "monkey debate" at the Oxford University Museum of Natural History on June 30, 1860. The great debate began with a two-hour-long address by Professor John William Draper of the Medical Department of New York University, invited as the major American champion of Darwinist thought. Thomas Huxley remembered Dr. Draper at the debate as "of course bringing in a reference to the *Origin of Species* which set the ball rolling." The details of what followed are controversial, but one exchange is engraved in the story of evolution.

Bishop Samuel Wilberforce—a premature televangelist and equivocal success as a mathematician—spoke next and taunted Huxley by asking if it was on his grandmother's or his grandfather's side that he was descended from apes. Huxley famously replied, "I asserted, and I repeat, that a man has no reason to be ashamed of having an ape for a grandfather. If there were an ancestor whom I should feel shame in recalling, it would rather be a man, a man of restless and versatile intellect, who, not content with an equivocal success in his own sphere of activity, plunges into scientific questions with which he had no real acquaintance, only to obscure them by an aimless rhetoric, and distract the attention of his hearers from the real point at issue by eloquent digressions and skilled appeals to religious prejudice."

Changeux's exhibit at Nancy displayed an earlier memento of a similar setback for the forces of unreason. This was the report of a royal commission appointed by Louis XVI to look into the activities of Franz Anton Mesmer (1734–1815). Mesmer had intruded his notion of "animal magnetism" into the highest level of French society. His doctrines leaned heavily on the Swedenborgian notion that matter is a subset of mind, a notion antithetical to the teachings of the philosophes and the French Academy itself. As Robert Darnton pointed out, there was a disturbing connection between the rise of mesmeric belief and the end of the Enlightenment in France. Mesmer taught that disease resulted from various obstacles to the flow of a magnetic "fluid" or vital energy in the body. In a mesmeric session, patients sat about in circular tubs and communicated the "fluid" by means of a rope looped about them all and by linking hands to form a mesmeric "chain." Soft music, played on wind instruments, a pianoforte, or a glass harmonica, reinforced the waves of ethereal energy that "entranced" the patient.

Reason struck back when the king appointed two commissions to investigate these practices. Dr. Guillotin (of the blade) headed one group of four prominent doctors from the Faculty of Medicine. The other commission was headed by Ambassador Benjamin Franklin (of the lightning) and boasted five members of the Academy of Sciences, including Bailly (of Jupiter) and Lavoisier (of oxygen). The commissioners spent weeks listening to descriptions of mesmeric theory and observing how its patients fell into fits and trances. They found false a report that being mesmerized through a door caused a woman patient to have a crisis. In Franklin's garden at Passy, a "sensitive" patient was led to each of five trees, one of which had been mesmerized. As the chap hugged each in turn to receive the vital fluid, he fainted at the foot of the wrong one. At Lavoisier's house, four normal cups of water were held before a mesmerized woman; the fourth cup produced convulsions, yet she calmly swallowed the mesmerized contents of a fifth, which she believed to be plain water. The commissioners concluded that there was no vital fluid. I'm reminded of Danny Kaye chanting in *The Court Jester,* "The pellet with the poison's in the vessel with the pestle; the chalice from the palace has the brew that is true!"

The verdict at Dover reminds us that the facts of evolution, no less than the laws of chemistry and physics, are the brew that is true.

10.

Spinal Irritation and the Failure of Nerve

In November 1864, the autumn between Gettysburg and Appomattox, Dr. Oliver Wendell Holmes, Parkman Professor of Anatomy and Physiology at Harvard Medical School, traveled to New York by train. He was accompanied on his journey from Boston by Julia Ward Howe, then at the peak of her fame for "The Battle Hymn of the Republic." Mrs. Howe recalled that they did not stop talking for the entire journey and that never had she been more vastly entertained. Holmes and she were coming to the Century Association to read appropriate verses at the seventieth birthday celebration of William Cullen Bryant. All three, Holmes, Howe, and Bryant, were of broad sanguine temperament and good humor; they were also old friends. Bryant was not only an abolitionist and an enthusiast of Central Park but also a celebrated man of letters who edited the *New York Evening Post.* And he was the lay leader of the Homeopathic Society.

Holmes was remarkably cordial that evening at the Century and recited his usual quota of immediately amusing, if easily forgettable, quatrains. The next day's accounts of the occasion do not mention whether Bryant remembered that Dr. Holmes was the leading opponent of homeopathic practice in the United States. In Holmes's pamphlet *Homeopathy and Its Kindred Delusions* (1842) he had referred to Bryant's homeopathy as "a mingled mass of perverse ingenuity, of tinsel erudition, imbecile credulity, and of artful misrepresentation." But the evening went well, presumably because a good dinner and a goodlier number of drinks can cause the lion to lie down with the lamb, even if the lamb won't get a good night's sleep.

Holmes had become Bryant's equal in literary renown by 1864; for almost a decade he had been featured in the *Atlantic Monthly,* America's most prestigious journal of arts and letters, where he had ample opportunity to temper his wisdom with a smile. He used almost every genre of humor to fill his monthly pieces, collected in *The Autocrat of the Breakfast-Table,* resorting often to the lowliest of them all, the pun. Holmes had himself complained in *The Autocrat* that "people who make puns are like wanton boys that put coppers on the railroad tracks. They amuse themselves and other children, but their little trick may upset a freight train of conversation for the sake of a battered witticism." (His own puns seem battered indeed. He had few regrets that his ancient family home in Cambridge was to be razed to provide new buildings for Harvard. He called it a case of "justified homicide." He saluted John L. McAdam, inventor of the paving process, as one of the seven wonders of the modern world: "The Colossus of Roads.")

But his sanguine spirit served him well during the years of war. Despite his unhappy trip to the field to collect his wounded son, despite the deaths of many of his former students, he remained a convinced patriot and propagandist for the Union cause. Lincoln's Democratic opponents thought the doctor a bit laughable, and his son was sometimes embarrassed by his father's armchair militancy. On September 13, 1863, Henry Livermore Abbot, a Copperhead comrade of young Holmes, reported home his battlefield opinion that young Captain Holmes "is a student rather than a man of action." He added, "His father, of course, one can't help despising." That sentiment was not surprising in a family of Copperheads. One week after the Emancipation Proclamation, Abbot had written home, "The president's proclamation is of course received with universal disgust, particularly the part which enjoins officers to see that it is carried out." Dr. Holmes had earned such enemies well. He responded to them in his own good-natured way in *The Autocrat of the Breakfast-Table:*

> "If a fellow attacked my opinions in print, would I reply?" asks the Autocrat. "Not I. Do you think I don't understand what my friend, the Professor, long ago called the hydrostatic

> paradox controversy?"—which enigmatic phrase he explained thus: "If you had a bent tube, one arm of which was the size of a pipe-stem and the other big enough to hold the ocean, water would stand at the same height in one as in the other. Thus discussion equalizes fools and wise men in the same way, and the fools know it."

The Autocrat's sanguine temperament also protected him from the Jamesian postwar failure of nerve. He remained as skeptical as he'd been before Fort Sumter and maintained the upbeat conviction that folly would by and large yield to reason. When next he came to New York he took aim at an opponent that had not surrendered at Appomattox: quackery. Dr. Holmes appeared before the graduating class of Bellevue Hospital Medical College in 1871. He began his talk by explaining the difference between the junior and senior members of the profession: the young doctor knows the rules, the older doctor knows the exceptions. He went on to warn the young graduates against the nostrums and "specifics" that passed for therapy in their century. Holmes's next targets were homeopathy, bleeding, and cupping. He advised the young physician to beware the homeopath and his clients:

> Some of you will probably be more or less troubled by that parody of medieval theology which finds its dogma in the doctrine of homeopathy, its miracle of transubstantiation in the mystery of its dilutions, its church in the people who have mistaken their century, and its priests in those who have mistaken their calling. You can do little with persons who are disposed to accept these curious medical superstitions. There are those whose minds are satisfied with the million-fold dilution of a scientific proof. No wonder they believe in the efficacy of a similar attenuation of herbs or potions. You have no fulcrum you can rest upon to lift an error out of such minds as these, often highly endowed with knowledge and talent, sometimes with genius, but commonly richer in the imaginative than the observing and reasoning faculties.

Holmes's major message was a stern warning that heroic measures, overdosing, and mineral purges were foolish ways of treating disease. He told the students of a surefire cure for agues and rheumatism, a cure safer than purges or bleeding. He advised them to pare the patient's nails, put the parings in a little bag, hang the bag around the neck of a live eel, and place him in a tub of water. "The eel will die and the patient will recover." This was an extension of the message Holmes had been reading to medical audiences since his manifesto at the Massachusetts Medical Society in 1860: "Throw out opium, which the Creator himself seems to prescribe . . . throw out wine, which is a food, and the vapors which produce the miracle of anesthesia, and I firmly believe that if the whole materia medica, as now used, could be sunk to the bottom—it would be the better for mankind and all the worse for the fishes."

Ever since doctors raised blisters, plied cauteries, and overdosed cases of what they called "spinal irritation" with calomel, the compliance of foolish patients has been exceeded only by the zeal of foolish physicians. Overuse of calomel, the cortisone of the nineteenth century, had led Holmes to believe that the French were in advance of the English and Americans in the art of prescribing for the sick without hurting them. He far preferred their various tisanes and syrups to what he called the mineral regimen of bug poison and ratsbane so long in favor on the other side of the Channel, much as he preferred French cuisine to the "rude cookery of those hard-feeding and much-dosing islanders." He asked whether calomel was not sometimes given by a physician on the same principle by which a landlord occasionally prescribes bacon and eggs—because he could not think of anything else quite so handy.

If calomel was the most overprescribed drug of the nineteenth century, spinal irritation was its most overdiagnosed ailment. "Some shrewd old doctors," Dr. Holmes told the Bellevue students, "have a few phrases always on hand for patients who will insist on knowing the pathology of their complaints. . . . I have known the term 'spinal irritation' to serve well on such occasions." It had already served well for over half a century. The spine became a locus for general malaise in 1821. Dr. R. P. Player of Mansbury reported that when he pressed on certain vertebrae, patients complained of pain and were often surprised at the discovery of tenderness in one or another bodily part: doctors and patients were puzzled. Along came the first complete account of spinal irritation in 1828, given by one Dr. Thomas Brown of Glasgow. He noted that while the spine was at the root of the problem, so to speak, the symptoms were often expressed elsewhere. Some of Dr. Brown's female patients displayed painful, tender spots (beneath the breast or under the sternum) of which they had not been aware until the examination. The morbid sensibility was chiefly in the skin: "The patient for the most part flinches more when the skin is even slightly pinched than when pressure is made on the vertebrae themselves. The pain is in the majority of cases more severe than in those of real vertebral diseases."

Spinal irritation crossed the Atlantic quicker than the submarine cable. At one time or another, most of the eminent Bostonians had their turn. As the Yankee men went off to their countinghouses, the malady became practically epidemic among the city's well-off ladies. The generic "madwoman in the attic" of nineteenth-century literature was given calomel or laudanum for spinal trouble. She tended to stay in the attic or a daybed for life, or until the end of the book. Spinal irritation was the excuse for ever more heroic therapy on the part of medical practitioners. Holmes reminded the young students that persons who seek the aid of a physician are very honest and sincere in their wish to get rid of their complaints, because they want to live as long as they can. But since they are desperate to stay alive at any cost, they accept that "heroic therapy."

The great majority of patients subjected to singeing and scalding, to cupping and leeching for aches and pains of the spine, were women. Their folly persists today: a review of recent textbooks and journal articles documents that women constitute between 85 and 95 percent of sufferers from the same symptoms that were once ascribed to "spinal irritations"; in the twenty-first century the ailments are called fibromyalgia, total food allergy, and chronic fatigue syndrome. A history of what passes as psychosomatic disease confirms the notion, tawdry but true, that women in pain are everywhere hassled by men with promises. One need not be a feminist critic to trace the foolish practices of today to their roots in medical misogyny.

History teaches us that social norms shape medical fashion and that medical fashion in turn shapes the symptoms that patients select. A prominent historian of medicine, Edward Shorter, suggests in his *From Paralysis to Fatigue* (1992) that "most of the symptoms of psychosomatic disease have always been known to Western society, although they have occurred at different times with different frequencies: Society does not invent symptoms; it retrieves them from the symptom pool."

As was the case with spinal irritation, many patients with fibromyalgia or chronic fatigue syndrome are no doubt afflicted by an as yet obscure response to chemicals or viruses; new candidates propose themselves weekly. But real microbes eventually cause diseases that really hurt, that kill, maim, or provoke detectable bodily ill. That's not the point of the psychosomatic argument. "The unconscious mind," our historian tells us, "desires to be taken seriously and not be ridiculed. It will therefore strive to present symptoms that always seem, to the surrounding culture, legitimate evidence of organic disease." The patient must present symptoms that the medical culture cannot reject. By hook or by crook, by stealth or by wealth, new symptoms arise that evade the doctor's new gadgets.

The diagnosis of spinal irritation and its treatment by "counterirritation" reached its zenith in *Spinal Irritation* (1886) by Dr. William Alexander Hammond of New York. He routinely found "multiple tender spots" in women suffering from spinal irritation and ascribed some of the cases to sexual excess or, so he was convinced, masturbation. In keeping with his contemporaries, he advocated treatment with "counterirritants" such as dry heat, scalding water, or croton oil extracts. Hammond had been for a short time surgeon general of the United States. He attained the distinction, unique for that rank, of being court-martialed. His trial occurred during the Civil War after acrimonious squabbles with Secretary of War Edward Stanton and bureaucrats of the career Medical Service. His chief problem, aside from a personal lapse in petty finance, seems to have been that he was too closely affiliated with the reformers of the Sanitary Commission.

But Hammond was able to recover from political infamy; eventually his reputation was cleansed and his rank restored by the Senate. He went on to become one of the founders of the American Neurological Association and also of the NYU Postgraduate Medical School, in the library of which his books are now quietly disintegrating. Hammond's magisterial *A Treatise on Diseases of the Nervous System* (1871), the first American textbook of neurology, is an extensive tome, published a scant seven years after his fall from official grace. The treatise is filled with outmoded rituals and jawbreaking syndromes. Scattered among this dross are neatly described case histories and new observations, but the volume is tough slogging.

In contrast, Hammond's other professorial book, *A Treatise on Insanity in Its Medical Relations* (1883), is only 738 pages long and can be read like fiction—the stuff of Maupassant or William Dean Howells. The chapters on "hysteria" and "hypochondriacal mania" yield the richest lode of stories. Hammond's characters include Eliza C., "who crowned herself with flowers, took a guitar, and announced that she was going to travel through the world. She got up in the night and washed her clothes in the chamber-pot. Then she had convulsions, mewed like a cat, tried to climb up the wall and finally fell into a state of stupor." Brief, mad lives are recounted by the hundreds in this astounding compendium, including the intervention of astute Colonel Charles May of the U.S. Army, who—by means of bloodless, bogus surgery—cured a fellow officer "of the belief that he was inhabited by chicken bones."

Poor Miss A.W., a patient of Hammond's, had the habit of swallowing pins by the dozen. She later extruded these pins from her skin, her nose, and various nether orifices (surely the first documented case of pins envy). We also learn of a "young lady from a Western city" who had "grasped a large knife that lay on the table and would have killed her mother with it had it not been seized by her sister, who was present." She explained this subsequently by saying that she had thought her mother "was a black man who was stealing her jewelry."

Hammond shows warmth and empathy for his mentally ill patients, regarding them in the tolerant, bemused fashion of a Victorian author displaying his fictional creatures:

> Kindness and forbearance, supported by firmness, will not altogether fail in their influence with even the most confirmed and degraded lunatics. Probably the most difficult class of patients to manage by moral means is that of the reasoning maniacs, and next to them those cases of hysterical mania which exhibit marked perversities of character and disposition. But even with such people the principles of justice and fair dealing will not be lost, and eventually an impression will probably be made on subjects incapable of being touched by other measures.

But woe to the patient who failed to respond to "moral means," the talking cure. She—for it was usually a female patient—was in for leeches or counterirritants:

> On the other hand, local bloodletting by cups or leeches is often a useful measure, especially in those cases in which there are pain and heat in the head accompanied with insomnia and excitement. A couple of leeches to the inside of the nostrils are remarkably efficacious in relieving cerebral hyperaemia [too much blood in the brain] and mitigating the violence of the physical and mental symptoms resulting from it. As to counter-irritants, such as blisters, croton oil, tartarized antimony, and the actual cautery, cases every now and then appear in which they seem to be of service. I have, however, several times aggravated the mental and physical symptoms of insanity by their use. I suppose the most generally advantageous agent of the kind is the actual cautery very lightly applied to the nucha [the nape] of the neck, but then the action in such a case can scarcely be called counter-irritant.

For simple hysteria, talking cures seem to work best, and it mattered not which learned profession did the talking or listening. A young lady of 20 began to wear "pads over the abdomen and gradually increased their thickness" until, confronted by her parents, she confessed to unwed pregnancy and declared a gentleman lawyer they all knew and respected to be her seducer. The outraged father offered the lawyer the alternative of "an immediate marriage or instant death from a pistol pointed at his head . . . the pistol, cocked, was very near his brain." Only after he had unwillingly agreed to this handgun marriage did the lawyer have a chance to chat with his betrothed. Using measures less severe than leeches or cautery, the lawyer by his tact and the directness of his questions succeeded in exposing the fraud and obtaining a full confession. The marriage was canceled and Hammond was engaged to complete the cure.

Some of these vignettes can be read as drafts for the melodramatic or gothic novels of Hammond's day; in this context, Dr. Holmes's *Elsie Venner* comes to mind. But the anecdotes were intended to reveal neither the character of individuals nor the nature of the society in which they moved. Instead, only enough detail was given to illustrate the variegated clinical forms of hysteria or spinal irritation.

Spinal irritation or inflammation eventually became the basis of an organized religion: Christian Science. Mary Baker Glover Patterson Eddy suffered a small epiphany when she hurt her back in 1866. Her injury was an obscure hurt that seemed to mimic an injury received by Henry James the younger in 1861. She was literally struck by the notion that since "matter and death are mortal illusions," one could overcome disease by exercise of mind. There is, in fact, a large area of agreement between this notion and those expressed by Henry James the elder in *Shadow and Substance.* What a Bostonian affliction! Poor Mary Baker Eddy was troubled all her life by "spinal inflammation and its train of suffering—gastric and bilious" to the point where her second husband had to carry her downstairs for her wedding ceremony—and back to her invalid bed directly thereafter. Sure enough, with the help of healing and the mind, she was soon able to climb all 182 steps of Portland's city hall tower. Eddy and her followers were persuaded that Christian Science and its healers constituted the main line of defense against "malicious animal magnetism," which was the main cause of illness and death. This not unpersuasive system of alternative medicine has continued to outlast its origins in spinal irritation.

As the last century wound down, rational France was again swept by medical unreason. Spinal irritation yielded to "hysteria" under the careful ministrations of Jean-Martin Charcot (1825–1893). Charcot began his career as an astute internist, and his clinical observations remain part of the literature of modern medicine. But once he turned to diseases of the mind, he became a Barnum of the mental wards. These days he is regarded as little more than a link between Mesmer and Freud. As with the ideas of Louis Agassiz, most observers agree in retrospect that there was more humbug than matter to Charcot's theory; if his talents were clinical, his genius was theatrical. His clinical demonstrations at the Salpêtrière hospital were open to the public, and a generation of dazed, sick women was displayed to an audience of fashionable voyeurs. The sessions were in stiff competition with the boulevard theaters of the *fin de siècle,* and perhaps for that reason most of the clinical syndromes Charcot described, such as *grande hystérie,* are now believed to have been either fancied or staged.

A student of Charcot, Dr. Jean-Albert Pitres of Bordeaux, found that attacks of *grande hystérie* took their origin in—you might have guessed—certain tender spots. In cases of hysteria, Dr. Pitres was able to define a "demarcated spasmogenic zone" over his patients' bodies (the small of the back, both armpits, and the bottom of the sternum) as well as over the ovary.

Sigmund Freud, who had traveled to the Salpêtrière to sit at Charcot's feet, was also a devotee of the tender-spot doctrine. He described hysteria in terms that might astound not only feminists. His patient Fraulein Elizabeth von R. had such an area, and when Freud touched a sensitive spot on her skin, she cried out. He could not help thinking that it was as though she were "having a voluptuous tickling sensation; her face flushed, she threw back her head and shut her eyes and her body bent backwards." In 1895, another of Freud's patients, Frau K., had cramp-like pains in her chest. "In her case," Freud told his friend Wilhelm Fliess, "I have invented a strange therapy of my own: I search for sensitive areas, press on them, and thus provoke fits of shaking which free her." His patient's spasmogenic zones, originally in her face, shifted to two points on her left chest wall—identical, Freud said, to his own spasmogenic points. But before we rush Freud to the wall for what we know to be the *real* meaning of all this, we might recall that the method we use to arrive at our conclusions was devised by him.

The remedy for all those myalgias, for all that accumulated pain and fatigue, was a century-long concentrated attack by male doctors on the female reproductive tract. More than half of the articles dealing with oophorectomy in 1889 described it as a treatment for mental disease; hysteria (from the Greek word for "womb") has, of course, been traditionally attributed to disorders of the womb or ovaries. Other modes of therapy for spinal irritation or hysteria included cauterization of the clitoris, hysterectomy, curettage, cupping, electroshock, and arsenic.

But the diagnostic fancies of pressure points and tender spots followed by counterirritants and nostrums are not limited to the past. The folly persists today, as can be seen from the fibromyalgia literature. The presence of eleven or more of eighteen "specific tender point sites" together with widespread pain in the *absence* of radiographic or lab abnormalities suffices nowadays for the official classification of fibromyalgia. Doctors who believe in this diagnosis today elicit pain at the magic spots of their mainly female patients by means of either a pain-metering machine or "palpation with the pulp of the thumb or the first 2 or 3 fingers at a pressure of approximately 4 kilos." One textbook expert of the 1990s suggests a variety of treatments for this syndrome under the general rubric "Counterirritant Therapy." These include "a wide variety of popular or unusual therapies, including massage, heat, liniments, steroid injections, ethyl chloride spray, and acupuncture," all of which he believes to be helpful in relieving pain and which sound very much like Hammond's prescriptions for spinal irritation. Since the magic potions haven't changed for over a century we might call this cult medicine rather than science. Our expert confirms that notion: "Factors influencing a choice of counterirritant technique will include simplicity, safety, and availability and economy on one hand and maximal placebo effect on the other."

Women, and not men, will be subject to these measures for the therapy not only of fibromyalgia but also "irritable bowel" and "chronic fatigue syndrome," the official criteria for which differ very slightly from those for fibromyalgia. One might therefore ask whether sexism, historical folly, or medical science is at work; homeopathy and its kindred delusions revisited, we might answer.

The world of Holmes and the black-bag doctors whom he addressed seems far away, viewed from our privileged decade of molecular medicine, of sono- and angiograms, of cyclo- and cephalosporins, of liposomes and liposuction. The day-to-day medicine from the 1830s to the 1930s appears to have been futile, groping—let's face it, *quackish.* All that laying on of hands, thumping of the chest, calomel, glycerine, cautery, compresses! Nevertheless, Holmes gave the young Bellevue physicians a meliorist vision of the medical profession that transcends the fashionable diagnoses and drugs of the day; it will remain true when colonoscopy will have gone the way of cautery, and cortisone the way of calomel, when he said:

> If the cinchona trees all died out . . . and the arsenic mines were exhausted, and the sulphur regions were burned up, if every drug from the vegetable, animal, and mineral kingdom were to disappear from the market, a body of enlightened men, organized as a distinct profession, would be required just as much as now, and respected and trusted as now, whose province should be to guard against the causes of disease, to eliminate them if possible when still present, to order all the conditions of the patient so as to favor the efforts of the system to right itself, and to give those predictions of the course of disease which only experience can warrant, and which in so many cases relieve the exaggerated fears of sufferers and their friends, or warn them in season of impending danger. Great as the loss would be if certain active remedies could no longer be obtained, it would leave the medical profession the most essential part of its duties, and all, and more than all, its present share of *honor.*

11.
Galton's Prayer

> What a tremendous stir-up your excellent article on prayer has made in England and America!
>
> —Charles Darwin to Francis Galton
> (November 8, 1872)

As the earth warms, hurricanes, typhoons, and tsunamis happen. After each such calamity, be it in the land of Muslim, Jew, or Gentile, our leaders launch worldwide appeals for material aid and urge public prayer for the injured. New research shows that they needn't bother with prayer.

Well over a century ago, Sir Francis Galton, Fellow of the Royal Society, a cousin of Charles Darwin and the founder of modern biostatistics, called for a prospective, controlled study of whether those for whom prayers were offered would heal faster than those unaided by distant appeals to the deity. Writing in the *Fortnightly Review* of August 1, 1872, Galton proposed the comparison of two groups of traumatically injured patients,

> The one consisting of markedly religious, piously-befriended individuals, the other of those who were remarkably cold-hearted and neglected [since] an honest comparison of their respective periods of treatment and the results would manifest a distinct proof of the efficacy of prayer.

In 2006, 134 years later, his call was answered in a definitive, double-blind study published in the *American Heart Journal*, fetchingly named "The Study of Therapeutic Effects of Intercessory Prayer (STEP)." Both its scope and cost were greater than the trial proposed in 1872.

Herbert Benson of Harvard and a brigade of faithful collaborators assigned three Christian prayer groups to pray for 1,800 patients undergoing coronary artery bypass graft (CABG) surgery in six medical centers throughout the United States. Funded mainly by the John Templeton Foundation, which supports research at the interface of religion and science, the $2.4 million study was described in the *New York Times* as "the most scientifically rigorous investigation of whether prayer can heal illness." It found that patients undergoing CABG surgery did no better when prayed for by strangers at a distance to them ("intercessory prayer") than those who received no prayers. But 59 percent of those patients who were told they were definitely being prayed for developed complications compared with 52 percent of those who had been told it was just a possibility, a statistically significant, if theologically disappointing, result. Benson et al. came to the objective conclusion that "Intercessory prayer itself had no effect on complication-free recovery from CABG, but certainty of receiving intercessory prayer was associated with a higher incidence of complications."

American advocates of intercessory prayer immediately raised objections to the study. "'God Factor' Defended; Prayer Study Flawed," headlined the *Worcester Telegram & Gazette*. Brother Dennis Anthony Wyrzykowski told the newspaper, "The study was not a ploy to make God look bad. . . . Dr. Benson is interested in the God factor. He's not out to disprove anything. We were disappointed. I was sure it would show that prayer works. . . ." "Area Residents Challenging Prayer Study," headlined the *Richmond Times-Dispatch*. The Rev. Robert Friend of All Saints Episcopal Church in Richmond told the paper, "That's not the way prayer works. When we pray we are aligning ourselves the best we can with God's will for us. I think God's will for us is that we be whole and healthy." All Saints would continue its weekly prayer service for healing.

Newspapers in less devout corners of the earth drew different conclusions: "Healing Power of Prayer Debunked" cried the *Gazette* of Montreal, while *Le Monde* warned from Paris that "*La prière serait dangereuse pour la santé*" ["Prayer Could be Dangerous to Your Health"]. Britain's *Guardian* called attention to the most unexpected result of the STEP study: "If You Want to Get Better—Don't Say a Little Prayer." The *Guardian*'s Oliver Burkeman wrote, "If a religious person offers to pray for you next time you fall ill, you may wish politely to ask them not to bother. The largest scientific study into the health effects of prayer seems to suggest it may make matters worse."

THE *Guardian* DIDN'T NEED BENSON'S STUDY to draw its conclusions. Indeed, the STEP trial had been scooped by an older, irrefutable, and far less expensive analysis, Galton's essay of 1872. Galton suggested that in place of any prospective study of intercessory prayer in traumatic injury, a more rigorous test of the effect of prayer would require life or death as an endpoint. He proposed an inquiry

> into the longevity of persons whose lives are prayed for; also that of the praying classes generally; and in both those cases we can easily obtain statistical facts.

He noted that, for hundreds of years, public prayer each Sunday had been offered from the Book of Common Prayer for the long life of the monarch of England, as in "Grant him [her] in health long to live." He asked, "Now, as a simple matter of fact, has this prayer any efficacy?" Fetching data from the records of over 6,500 biographies assembled by a colleague, Galton found that intercessory prayer offered on behalf of those who knew they were prayed for was bad for one's health.

It turned out that members of royal houses had a mean life expectancy of 64.0 years, significantly less than that of other aristocrats (67.3 years) or of other gentry (70.2 years). Since the royals knew they were being prayed for—we might say that each Sunday they were on the same page as their subjects—they were as much at risk for a worse outcome from prayer as were Benson's CABG patients. While Benson et al. downplayed the possible side effects of prayer as a possible chance finding, Galton had reached a similar conclusion from a neater endpoint.

Worse yet! The life expectancy of eminent clergymen (66.4 years) was no higher than that of their peers among lawyers (66.5 years) or medical men (67.0 years), and Galton concluded that, in keeping with the results of the STEP trial,

> prayers of the clergy for protection against the perils and dangers of the night, for protection during the day, and for recovery from sickness, appear to be futile in result.

In an editorial in the *American Heart Journal* that accompanied the STEP trial paper, William Krucoff and colleagues at Duke sounded concern over the higher incidence of complications in the group that knew it was being prayed for. They asked: "If the results had shown benefit rather than harm, would we have read the investigators' conclusion that this effect 'may have been a chance finding,' with absolutely no other comments, insight, or even speculation?" Supported by a foundation committed to the "God factor" at work in health and disease, Benson et al. must have been taken aback by the harmful consequences of intercessory prayer. Perhaps that's why it took almost five years to analyze data obtained on patients enrolled in the trial from January 1998 to November 2000! Undaunted, Dr. Charles Bethea, one of STEP's coauthors, insisted to the *New York Times* in 2006 that "One conclusion from this is that the role of awareness of prayer should be studied further." While the final STEP publication was impartial in its presentation and low-keyed in interpretation, it aroused the anger of believers and skeptics alike. The credulous contended that STEP was flawed, that it represented bad medical care and trivialized religion. Skeptics argued against the study, convinced that there is no place in the realm of science for supernatural intervention.

Now it is certainly within the prerogative of objective clinicians to engage in statistical analyses of long-range prayer for others, especially when a foundation devoted to such notions picks up the tab. On the other hand, not only skeptics will wonder why the National Institutes of Health would encourage or support inquiries into the supernatural. Newspaper accounts of the STEP trial carried the remarkable news that our government has spent more than $2.3 million of public money on prayer research since 2000. Some of these studies overlap the published results of Galton and Benson et al., including, for example, "Distant Healing Efforts for AIDS by Nurses and 'Healers'" (Elisabeth F. Targ, Principal Investigator California Pacific Medical Center, 1-R01-AT-485-1), a three-year grant totaling approximately $663,000, and "Efficacy of Distant Healing in Glioblastoma Treatment" (Elisabeth F. Targ, Principal Investigator, 1-R01-AT-644-1), a four-year grant totaling approximately $823,000. Sadly, the results of these studies remain largely unpublished: Dr. Targ died of glioblastoma in 2002. In April 2000, she had reported:

> Of more than 135 studies of distant healing on biological organisms . . . about two-thirds reported significant results. One fascinating study . . . concerned remote healing of tumors on mice. The study showed that the healers who were farthest from the mice had the greatest influence in shrinking the tumors.

To be fair, when conducting controlled trials, Dr. Targ had been as professional and objective as Dr. Benson and his colleagues in their STEP study. Reporting on an $800,000 trial funded by the Department of Defense (Grant No. 17-96-1-6260) of Complementary and Alternative Methodologies (CAM) in breast cancer, "The Efficacy of a Mind-Body-Spirit Group for Women with Breast Cancer," she concluded that "The study found equivalence on most psychosocial outcomes between the two interventions" (CAM versus control). And that's not even at a distance.

This year, the chances of being funded on any given grant application to the NIH are around 15% percent. It is in this context that one questions whether the NIH, and especially the National Center for Complementary and Alternative Medicine, NCCAM (now the National Center for Complementary and Integrative Health, NCCIH), has any business encouraging further grant applications and/or research into prayer. Those of us who engage in experimental biology are generally uninterested in enlarging the norms of our realm into the spiritual, artistic, or ethical life of our time. But believers in "noetic," spiritual, or supernatural explanations for the vast territory of the unknown in science seem to have no such qualms. They've persuaded a credulous citizenry that there is spiritual gold to be mined by applying the methods of science to the study of religious practice. By confusing credulity with piety, they've also cleared the way to belief in "intelligent design." While such notions discredit both rigorous science and true beliefs, they are part and parcel of the new Endarkenment.

NCCAM was undeterred. In 2005 Catherine Stoney, PhD, of its Division of Extramural Research and Training, insisted that "There is already some preliminary evidence for a connection between prayer and related practices and health outcomes. For example, we've seen some evidence that religious affiliation and religious practices are associated with health and mortality—in other words, with better health and longer life." She is unlikely to have consulted Galton's statistics in the *Fortnightly Review*.

Galton, who shared a grandfather—Erasmus Darwin—with Charles Darwin (all three were Fellows of the Royal Society), was far more modest in his peroration, as he offered the skeptical equivalent of prayer:

> Neither does anything I have said profess to throw light on the question of how far it is possible for man to commune in his heart with God . . . and it is equally certain that similar benefits are not excluded from those who on conscientious grounds are skeptical as to the reality of a power of communion. . . . They know that they are descended from an endless past, that they have a brotherhood with all that is, and have each his own share of responsibility in the parentage of an endless future. The effort to familiarize the imagination with this great idea has much in common with the effort of communing with a God, and its reaction on the mind of the thinker is in many important respects the same. It may not equally rejoice the heart, but it is quite as powerful in ennobling the resolves, and it is found to give serenity during life and in the shadow of approaching death.

12.

Dr. Doyle and the Case of the Guilty Gene

> The importance of the infinitely little is incalculable. Poison a well at Mecca with the cholera bacillus, and the holy water which the pilgrims carry off in their bottles will infect a continent, and the rags of the victims of the plague will terrify every seaport in Christendom.
>
> —Dr. Joseph Bell (1892)

EDGAR ALLAN POE MAY HAVE WRITTEN the first mystery, but Dr. Arthur Conan Doyle put the genre on the map for keeps. Those ever-popular stories written by Doyle and Poe share a pattern: the hero is an omniscient, eccentric detective; the narrator is his obtuse roommate; the police are severely befuddled; the solution hinges on recourse to natural history or the pharmacopoeia. Most of all they celebrate what Poe called "the mental features discoursed of as the analytical." However, despite the similarities in their fictions, Poe and Doyle were very unlike with respect to fortune, temperament, and career. They differed as much from each other as, let us say, doctor and patient. Doyle went as a medic to the Boer War, while "poor Edgar" died in a hospital of drink and tuberculosis. They're also as different as checkers and chess.

The Murders in the Rue Morgue begins with a curious introduction, a mini-essay in which Poe contrasts the "analytic powers" needed for chess and checkers. He argues that the mental features required in "the elaborate frivolity of chess" rank on a lower scale than those wanted to solve the "unostentatious" game of draughts (checkers). In chess, he claims, "the attention is here called powerfully into play . . . it is the more concentrative rather than the more acute player who conquers," whereas in the simpler game of checkers, "what advantages are obtained by either party are obtained by superior *acumen* . . . by some *recherché* movement." Reading Poe's argument in the context of his rejection by the literary lions of his day, we have a hint of what he was up to when he proposed:

> Between ingenuity and the analytic ability there exists a difference far greater indeed than between the fancy and the imagination but of a character very strictly analogous. It will be found, in fact, that the ingenious are always fanciful, and the truly imaginative never otherwise than analytic.

Since Poe accused the Boston literati, and especially the transcendentalists, of excess fancy and limited imaginations, we might read his homage to the "truly imaginative" as an advertisement for himself. Poe, champion of the recherché in his tales of terror and the imagination, seems to anticipate how critics of the future might distinguish his work from that of his disciple, Sir Arthur Conan Doyle. It's the difference between England and France, chess and checkers. The good doctor's very Anglo-Saxon hero, Sherlock Holmes, excels in the attentive power of a chess player. On the other hand, Poe's Auguste Dupin shows the acumen of a Gallic champion at checkers with his recherché move into Cuvier. "King me!" the story seems to be saying: "The ape did it!"

Poe's quest for whodunit, in which the murderer ranks among the highest of apes, suggests that the detective story had other overtones from its beginning. The murders in the rue Morgue were committed in those critical years for natural history between the theories of Jean-Baptiste Lamarck (1809) and those of Charles Darwin (1859). Baron Georges Cuvier's original encyclopedia of the animal kingdom showed European man standing first in the Great Chain of Being, followed closely by the great apes. As one might expect, a very Gallic man was depicted at the head of the human pack, with the other specimens (Asian, Semitic, African) trailing backward to the beasts. The apes were described as "almost human, but with distinctly violent and bestial features."

With Cuvier as a template, it may not be too fanciful to suggest that the murder mystery began as a search for the ape beneath the skin, the *biology* of social guilt. And so, notwithstanding the differences between attention and acumen, checkers and chess, that Poe used to explain why he invented the genre, one has a hunch that the detective story is out not only to find whodunit but also to search for the guilty gene.

More than a century ago, in December 1892, there appeared in the *Bookman* a review of Dr. Arthur Conan Doyle's recently published *The Adventures of Sherlock Holmes.* The piece was written by Dr. Joseph Bell, a professor of surgery at the University of Edinburgh. Readers of the *Bookman,* a popular literary magazine, knew that the Scottish surgeon was the real-life prototype of Dr. Doyle's detective—Doyle had been Bell's student and assistant—and could fancy that they were reading a review of Sherlock Holmes's adventures written by Sherlock Holmes himself. Bell wrote:

> The greatest stride that has been made of late years in preventive and diagnostic medicine is the recognition and differentiation by bacteriological research of those minute organisms which disseminate cholera and fever, tubercle and anthrax. The importance of the infinitely little is incalculable. Poison a well at Mecca with the cholera bacillus, and the holy water which the pilgrims carry off in their bottles will infect a continent, and the rags of the victims of the plague will terrify every seaport in Christendom. Trained as he has been to appreciate minute detail, Dr. Doyle saw how he could interest his intelligent readers by taking them into his confidence, and showing his mode of working. He created a shrewd, quick-sighted, inquisitive man, half doctor, half virtuoso, with plenty of spare time, a retentive memory for whom the petty results of environment, the sign-manuals of labor, the stains of trade, the incidents of

> travel, have living interest, as they tend to satisfy an insatiable, almost inhuman, because impersonal curiosity.

Dr. Bell might have been describing himself. But Bell was not the only model for Dr. Doyle's half doctor, half virtuoso. And although the records are murky, it should come as no surprise that Sherlock Holmes owes as much to the Autocrat of the Breakfast-Table as to the admirable Dr. Bell. Doyle made it clear in his memoirs that he borrowed "Dr. Watson" from the surname of a fellow practitioner in Portsmouth, a Dr. James Watson. But extensive searches of Doyle's memoirs, correspondence, and scrapbooks have yielded not a word as to the origin of the name "Holmes." Is there a secret? And if so, what does the secret tell us about Sir Arthur Conan Doyle and his search for the genetically guilty?

Dr. Bell tells us that a medical diagnostician and a detective share an imagination "capable of weaving a theory or piecing together a broken chain or unraveling a tangled clue." Well, it may be almost elementary—so to speak—to piece together a chain of evidence that traces the invention of Sherlock Holmes to the best-known doctor-writer of the nineteenth century: Dr. Oliver Wendell Holmes, the *Autocrat*, *Poet*, and *Professor at the Breakfast-Table*, to name but three of his volumes. Indeed, I would argue that the young Dr. Doyle not only appropriated Holmes's last name for his hero but also followed the Autocrat's path to literary fame by playing the patriot card. Unfortunately, he also carried the biological Darwinism of Holmes to its social extreme. Conan Doyle got into the business of looking for guilty genes early in his career.

In September 1892, when young Dr. Doyle was by no means a household name, it was announced in the press that the *Foudroyant,* Admiral Nelson's old flagship, had been sold to the Germans—and for scrap, at that! Doyle boiled over and dispatched these verses to the press as "A Humble Petition":

> Who says the Nation's purse is lean,
> Who fears for claim or bond or debt
> When all the glories that have been
> Are scheduled as a cash asset?

If times are black and trade is slack,
If coal and cotton fail at last,
We've something left to barter yet—
Our glorious past . . .

There's many a crypt in which lies hid
The dust of statesman or of king;
There's Shakespeare's home to raise a bid,
And Milton's house its price would bring.

What for the sword that Cromwell drew?
What for Prince Edward's coat of mail?
What for our Saxon Alfred's tomb?
They're all for sale!

What was it that prompted this outburst of poetry in a newspaper by a writer known to Dr. Bell and readers of the *Bookman* as "a born story teller" who wrote for magazines? It seems likely that the example, in both sentiment and meter, was that of Oliver Wendell Holmes. In September 1830 there had appeared a short notice in the *Boston Daily Advertiser* that the frigate *Constitution,* proud veteran of the American Navy, was about to be dismantled. Young Holmes—not yet a medical student, but already the class poet of Harvard '29—dashed off a poem to the *Advertiser.* That poem, "Old Ironsides," like Doyle's plea for the *Foudroyant,* was an indignant response to a tightwad regime. It gained for Holmes an immediate national reputation. Indeed, until television stamped out juvenile literacy on this continent, generations of American schoolchildren knew Holmes's poem by heart:

Ay, tear her tattered ensign down!
    Long has it waved on high,
And many an eye has danced to see
    That banner in the sky;
Beneath it rung the battle shout,
    And burst the cannon's roar;—

> The meteor of the ocean air
> Shall sweep the clouds no more.
>
> Her deck once red with heroes' blood,
> Where knelt the vanquished foe,
> When winds were hurrying o'er the flood
> And waves were white below,
> No more shall feel the victor's tread,
> Or know the conquered knee;—
> The harpies of the shore shall pluck
> The eagle of the sea!

The poem was so well known in America that several generations of medical students used to recite a parody that began "Ay, tear her tattered enzyme down . . . "

Ample evidence suggests that Doyle was a close reader of American writers: the late John Dickson Carr was not the only critic to note the strong influence of Bret Harte on the doctor's earliest adventure stories (see *The American's Tale,* 1879), and Poe's *Murders in the Rue Morgue* became Doyle's immediate example. Doyle's notebooks and the drafts for *A Study in Scarlet* clearly show that he turned his attention to a procedural puzzle after abandoning a murder tale inspired by Poe. In keeping with his Gallic model, Doyle was going to call that early effort *The Lerouge Case*; you'll recall that the hero of Poe's "The Gold Bug" is named Le Grand. But rouge turned scarlet, and Doyle's notebooks show him groping for a method: "The coat-sleeve, the trouser-knee, the callosities of the forefinger and thumb, the boot—any one of these might tell us, but that all united should fail to enlighten the trained observation is incredible."

Doyle knew that in order to turn clinical observation into a story he required not only a new method but also a new kind of character who might apply that method. He needed an observer, an experimentalist who would make criminal investigation an exact science. "By a study of minutiae, footprints, mud, dust, the use of chemistry and anatomy and geology, he must reconstruct the scene of a murder as though he had been there."

While Joseph Bell was a good enough model for a detective who could observe the minutiae of evidence, clinical observation was only one talent required of a "stoop-shouldered wizard of lens and microscope." No—Doyle needed someone of more quantitative bent, someone who might be scholar enough to write a monograph describing "one hundred and fourteen varieties of tobacco ash." For that model Doyle required a specialist: an anatomist, a microscopist, a physiologist. And who was more qualified to serve as that model than the Parkman Professor of Anatomy and Physiology at Harvard, the doctor-poet who had introduced histology to America, Dr. Oliver Wendell Holmes? There is little doubt that Dr. Holmes must have been very much on Dr. Doyle's mind in March and April 1886.

Sherlock Holmes makes his first appearance from behind a chemistry bench in St. Bartholomew's Hospital, where Dr. Watson mistakes him for a medical student—Holmes is working on a test for occult blood that will replace the time-tested guaiac method. The book in which the detective first appears is *A Study in Scarlet,* written in the spring of 1886. And in the spring of 1886 the life and times of one doctor were being celebrated in every British newspaper. Oliver Wendell Holmes ranked at the time with Emerson, Lowell, Longfellow, and Hawthorne in British esteem. He had arrived on a triumphant hundred-day tour of England, in the course of which he received honorary degrees from Oxford, Cambridge, and—oh, yes—Edinburgh. The doctor's doings on the London social scene were duly recorded; there were literary dinners with Robert Browning, Henry James, Walter Pater, and George du Maurier; state receptions with dukes and earls at the side of the prime minister (Gladstone); visits to the poet laureate (Tennyson) and to the Derby (with Prince Albert Edward); professional receptions where his companions were Sir James Paget (of the disease) and Sir William Gull (of the thyroid). Such were the joys—young Dr. Doyle must have noticed—of a successful career in literature and medical science.

Added to this celebrity was Holmes's association with one of the most notorious criminal trials of the century. Every English connoisseur of murder was aware of the Webster-Parkman case, the Harvard Medical School murder of 1849. As Simon Schama has reminded us in *Dead Certainties,* John Webster, the Erving Professor of Chemistry, was accused of having killed, dismembered, and almost destroyed the remains of Dr. George Parkman. Parkman, a wealthy practitioner and sharp businessman, had given the land on which Harvard Medical School stood, and the university rewarded him by naming the chair of anatomy and physiology after him. Dean Holmes, who occupied that chair, testified for the prosecution. As an expert witness he had confirmed that "a large mass of human bones, fused slag and cinders . . . the block of mineral teeth and the gold filling" found in the ovens of Professor Webster's chemistry laboratory were the remains of the unfortunate Parkman. These residues were also identified by Holmes's anatomical colleague Dr. Jeffries Wyman—he of the Philosophers Camp in the Adirondacks. Their careful analyses helped the prosecution to prove foul play and to uncover Dr. Webster's postmortem high jinks with the body. (One might say he was the first Harvard biochemist found guilty of cooking the data.) Detailed accounts of the trial in the English press never failed to mention Dr. Holmes. A few years after Webster was convicted and hanged, Charles Dickens, one of many English followers of that grisly crime, persuaded Dr. Holmes to walk him through Harvard's chemistry lab to view Webster's furnace, proof that every English literary gent knew that the Parkman Professor was privy to the secrets of the Parkman case.

We have ample reason, therefore, for supposing that Doyle conflated several aspects of Dr. Holmes in his detective hero. The name, certainly. The doctor-author model, probably. The forensic bent, surely. But the scientific urges? That numerical drive? Do these derive from Bell the surgeon or from Holmes the physician? Holmes, on the evidence, one might argue. Holmes had been trained in the quantitative methods of Pierre-Charles-Alexandre Louis and drilled in Louis's motto, "*Formez toujours les idées nettes. Fuyez les à peu près.*" Dr. Holmes paraphrased it as: "Always make sure that you form a distinct and clear idea of the matter you are considering. Always avoid vague approximations where exact estimates are possible; *about so many—about so much*—instead of the precise number and quantity." Those words might have dropped from the mouth of Sherlock Holmes when Watson lowered his newspaper, as he did in *The Resident Patient*, to look up at an unframed picture on the wall of Henry Ward Beecher (brother of Harriet Beecher Stowe, lifelong friend and correspondent of Dr. Holmes). They do, in fact, practically drop from the mouth of "young Stamford" in *A Study in Scarlet* as he describes Holmes to Watson: "He appears to have a passion for exact and definite knowledge." *Fuyez toujours le à peu près*, Stamford might have said.

Dr. Holmes returned from his studies in France fully aware of his responsibility for introducing new methods into a medical backwater: "I have lived among a great, a glorious people; I have thrown my thoughts into a new language; I have received the shock of new minds and new habits." Once at Harvard, he inaugurated the study of medical microscopy in the United States. He went on to design a number of teaching and research microscopes, inventing a portable microscope for use in the classroom. Following the lead of Samuel F. B. Morse, who had brought the daguerreotype back from his own Paris sojourn, Holmes became an ardent photographer. The best-known portraits of Dr. Holmes depict him in his study, posed in front of a microscope or the odd optical instrument. He was proudest of his achievements in histology, of what the achromatic lens had wrought, asking his Harvard students in 1861:

> Now what have we come to in our own day? In the first place, the minute structure of all the organs has been made out in the most satisfactory way. The special arrangements of the vessels and the ducts of all the glands, of the air-tubes and vesicles of the lungs, of the parts which make up the skin and other membranes, all the details of those complex parenchymatous organs which had confounded investigation so long, have been lifted out of the invisible into the sight of all observers . . . Everywhere we find cells, modified or unchanged. They roll in inconceivable multitudes (5 million or more to the cubic milliliter) as blood whisks through our vessels . . . they preside over the chemical processes which elaborate the living fluids [and] the soul itself sits on a throne of nucleated cells.

Professor Holmes throbbed with satisfaction at the new world under his lens. He did not foresee that, in less than a generation, the routine use of portable and powerful microscopes in the field would turn the sanitary revolution into the bacteriological revolution. He *did* anticipate Dr. Bell's tribute to the importance of the infinitely little. Indeed, the passage leads us to suspect that a lean sleuth with lens and microscope was exactly what a young medical writer *would* come up with who had been weaned on the teachings of Bell and the writings of Holmes. For Holmes was more than a sedentary professor of anatomy. As a physiologist he had moved to more kinetic studies. He had written scholarly and popular articles on the gait of man and beast: footprints on the sands of time. Lens in hand, he pointed out to his Harvard classes that anatomy studies the organism in space, whereas physiology studies it in time. And when the illustrator Sidney Paget drew the restless Sherlock Holmes, he often showed him before a microscope or with lens in hand. There, fixed on the pages of childhood, the lean sleuth will forever peer through the lens at the spoor in the mud, or exhort his companion, "Come, Watson, come! The game is afoot." Dr. Doyle took from Dr. Holmes's thirteen volumes of writing the lesson that experimental science could produce a popular literature. Private investigation could bring popular fame. And is it no accident that by the end of the nineteenth century, doctors who studied disease began to call themselves clinical *investigators* rather than clinical *observers*? Sure enough, by the time the American Society for Clinical Investigation was founded at the beginning of the twentieth century, both doctor and detective had become investigators; both used the microscope to study the importance of the infinitely little.

THE SHIFT FROM OBSERVATION TO INVESTIGATION also took place in the nineteenth-century novel. The literary scholar Lawrence Rothfield has observed in *Vital Signs* (1992) that the displacement of realism (Balzac) by naturalism (Zola) correlated in time and sensibility with the "displacement of one form of scientific thought (that of clinical medicine) by another (that of experimental medicine)." What Rothfield has called the "invasive aspects" of clinical and criminal investigation led him to explore the Freudian roots of detective fiction. He suggested that the pleasures of criminal investigation are almost erotic. Holmes sloshes through bog and moor to glimpse—at last—the primal scene. The detective, with nostrils dilated, cheeks flushed, hot after the scent of a criminal—or the literal denouement of the family romance—resembles in his rush of excitement the scientist who has snared an offending microbe: "Yes, I have found it!"

But Poe was always ahead of Doyle, even in the Freudian sweepstakes. Here is his portrait of the Chevalier Auguste Dupin in deductive rapture at the moment of discovery: "His manner at these moments was frigid and abstract; his eyes were vacant in expression; while his voice, usually a rich tenor, rose to a treble . . . " Poe precedes this orgasmic description with another that tells us as much about Watson and Holmes as about the narrator and Dupin: "He boasted to me . . . that most men, in respect to himself, wore windows in their bosoms, and was wont to follow up such assertions by direct and very startling proofs of his intimate knowledge of my own." Oliver Wendell Holmes wrote what he called his medicated novels to provide popular, sensational entertainment, and Rothfield could have had Holmes of Boston, rather than Doyle of Edinburgh, in mind when he describes the "sensational effects that detection can produce in the detected, the detective and the reader—effects that themselves are being given a scientific, even medical, status at the very moment when the detective story comes into its own as a genre: the moment of Holmes." But the moment of Holmes—which was also the moment of naturalism—had Darwinian as well as Freudian undertones. Doctor and detective alike used the tools of science to root out the abnormal in body or body politic. It is probably again no accident that Francis Galton (1822–1911), who founded the science of eugenics, developed fingerprinting by microscopy as part of his campaign of racial classification (1891). And eugenics turned out to be the applied science of social Darwinism. As Dr. Bell pointed out, "Racial peculiarities, hereditary tricks of manner, accent, occupation or the want of it, education, environment of all kinds, by their little trivial impressions gradually mold or carve the individual, and leave finger marks or chisel sores which the expert can recognize."

Here then was a new scientific method to find the ape beneath the skin, to search in the odd thumbprint for those racial peculiarities that only the expert can recognize! Galton himself makes an appearance in the penultimate paragraph of Dr. Bell's review. Galton's fingerprint method, Bell argued, renders "the ridges and furrows of the stain visible and permanent." Those stains were racial and indelible. No wonder that we nowadays call the technique used to identify DNA fragments by restriction enzymes (RFLP) "fingerprinting."

Informed by hindsight, we can accuse a literature based on Galtonian eugenics of social Darwinism. In *The Adventures of Sherlock Holmes* and its sequels, moral flaws are signified by physical abnormalities. Doyle offers us stock villains aplenty: the Gipsy, the Moor, the Levantine, the Man with the Blue Carbuncle. Those *others*. The detective ferrets out their guilt not from behavior alone but from Bell's "myriads of signs eloquent and instructive which need the educated eye to detect." Each twist of the spine, droop of a lid, bend in the nose, or blotch on the skin is an outward sign of a deeper flaw in the flesh. Those racial stains, which offend the physical standards of a settled race, are as visible and permanent as a fingerprint. The genes they express are the genes of guilt. Sherlock Holmes presided over the most thrilling gene hunt of his century and became a hero of adolescents worldwide. Meanwhile, Sir Arthur Conan Doyle became the staunch dean of empire loyalists and—like Yeats and William James before him—lapsed into spiritualism in old age. But as a student of Dr. Bell and fan of Dr. Holmes, Dr. Doyle put on the map forever the notion that doctors and detectives are after the same game: "Come,Watson, come! The game is afoot." It was the moment of Holmes indeed.

After the moment of Holmes, in fact as in fiction, it became the task of clinical and criminal investigators—of docs and cops—to root out the transgressor, to incriminate man or microbe. Edmund Wilson described what happens to a villain in the nineteenth-century detective story: "he has been caught by an infallible Power, the supercilious and omniscient detective, who knows exactly where to fix the guilt." We have learned over the last century that social Darwinists—detective or dictator—have had no qualms about fixing guilt on one or another gene. If this lesson smacks of anachronism, so be it.

The rise of the detective story in the years between Lamarck and Darwin is one aspect of the nineteenth-century argument in favor of nature over nurture. In that sense, murder is no mystery; when a detective scans the palm or the face he is reading the gene. Holmes's Elsie Venner, Doyle's Moriarty, even Poe's ourang-outang are cast as losers in the Darwinian lottery, examples of *others* who could no more change their nature than their serpentine eye or their furrow of skin. On the other hand, a meliorist would argue that if guilt lies in the gene, the guilty are more to be pitied than censured. That is exactly what Oliver Wendell Holmes argued in *Elsie Venner.* Those whom Dr. Doyle and the social Darwinists turned into villains, Dr. Holmes and the Boston meliorists regarded as victims. Snakebit in the womb but guiltless before God, they were more to be pitied than censured. What is given by nature can be forgiven by man. Following that example today, we might say that if "special influences" work on behavior like ferments in the blood, it is the task of our new genetics to give them any name but guilt.

Dr. Doyle may not have learned genetic sweetness and light from Dr. Holmes, but he remained forever in the Autocrat's debt for the name of his sleuth. On his first trip to America in the autumn of 1894, he visited Boston to view Mount Auburn Cemetery. Among the russet elms, yellow willows, and golden maples, he saw the graves of those who had presided over the Flowering of New England: "Lowell, Longfellow, Channing, Brooks, Agassiz, Parkman and very many more." And he did one other thing. "Yesterday," he wrote, "I visited Oliver Wendell Holmes's grave and I laid a big wreath on it." Why Holmes alone of all the others? Elementary, my dear Watson!

Two for the Road

13.

Swift-Boating "America the Beautiful": Katharine Lee Bates and a Boston Marriage

O beautiful for spacious skies
For amber waves of grain,
For purple mountain majesties
Above the fruited plain!
America! America!
God shed his grace on thee
And crown thy good with brotherhood
From sea to shining sea!

—Katharine Lee Bates, "America the Beautiful" (1893)

I strongly believe the neglected American people need . . . leadership and our Country needs to return to America The Beautiful in every way possible.

—Linda Archer, reader comment on washingtonpost.com (April 18, 2007)

In the summer of 2007, as political tempers flared in early skirmishes from Iowa to the Carolinas, one rather nasty theme emerged. "Iowa Gay

Marriage Ruling Stirs 2008 Race," ran the AP headline, and a contest began for the bipartisan laurels of bigotry. And sure enough, between mug shots at weenie roasts and platitudes at county fairs, a handful of hopefuls warned the faithful that marriage between people of the same sex ranked among the major threats to our republic.

Irony rampant: the same cameras that showed us politicians of every stripe and party disporting themselves at the Iowa State Fair also featured squeaky-clean farm kids welcoming visitors to the fair with the rousing verses of "America the Beautiful." So on behalf of a good number of my fellow citizens—and of their legislators as well—I'd like to remind both present and future candidates for office of Katharine Lee Bates, who wrote the poem "America the Beautiful." It's the story of a happy Boston marriage in the Era of the White City.

Katharine Lee Bates (1859–1929) is the most famous native of Falmouth, Massachusetts; her statue decorates the library lawn, the road to the library bears her name, the bicycle path along Vineyard Sound to Woods Hole is named "The Shining Sea," and the upscale granola store is called "Amber Waves." Her poem "America the Beautiful" is usually sung to music set by Samuel A. Ward, a Son of the American Revolution. It pays homage to their New England forebears:

> O beautiful for pilgrim feet,
> Whose stern, impassioned stress
> A thoroughfare for freedom beat
> Across the wilderness!
> America! America!
> God mend thine ev'ry flaw,
> Confirm thy soul in self control
> Thy liberty in law!

It's a sentiment less bellicose than that expressed in our official national anthem—"the bombs bursting in air"—and considerably sweeter than the boast of "Deutschland Über Alles," the pomp of "God Save the Queen," or the gore of "Le Marseillaise." It is also a fitting postbellum sequel to Julia Ward Howe's "The Battle Hymn of the Republic." Indeed, Julia and Samuel A. Ward were Yankee kin.

Bates was inspired to write "America the Beautiful" on her first trip out West. A professor of English at Wellesley, she had been asked to teach English religious drama at a summer school in Colorado Springs and, she recalled, "spent a happy three weeks or so under the purple range of the Rockies." To celebrate the end of the session, she and others on the faculty made an excursion to Pike's Peak, pulled to the summit in mule-drawn prairie wagons that bore the slogan "Pike's Peak or Bust!" She went on, "It was at the summit, as I was looking out over that sea-like expanse of fertile country spreading away so far under those ample skies, that the opening lines of the hymn floated into my mind."

She left Colorado Springs with notes for the entire four stanzas and other memorabilia of her extended trip to the Rockies, but the poem did not appear until July 4, 1895, when it was published in the *Congregationalist*. After a musical setting by the once well known Silas G. Pratt attracted national attention, the popular stanzas became open game for other musical versions, and by 1923 more than sixty "original" settings had been perpetrated. The verses can be sung to many old tunes, including "Auld Lang Syne" and "The Harp That Once through Tara's Halls." But the setting we know best nowadays was adopted by Ward from the hymn "Materna," and the words we use are those of Bates's revised version of 1913.

"THY LIBERTY IN LAW!" COULD SERVE AS A MOTTO for Bates and her impassioned generation of pilgrim daughters. Bates was graduated from Wellesley in 1880, ten years after that stern, seminary-style college had been chartered as a place for "noble, white, unselfish Christian Womanhood." But by 1882, winds of change from the West brought a new generation in the person of Alice Freeman (1855–1902). Freeman was only 27 when she was called from the coeducational University of Michigan to become Wellesley's second president. She proceeded to transform Wellesley into a college ready for the twentieth century; she also helped to found Radcliffe and the Marine Biological Laboratory.

As president of Boston's Woman's Education Association (WEA), Alice Freeman collaborated with Elizabeth Cary Agassiz, widow of biologist Louis, to work out the legal arrangements whereby the Harvard Annex for Women became Radcliffe College. The WEA also raised $10,000 to promote the teaching and research by women in science. This gift made it possible in 1888 to purchase land near the all-male U.S. Fisheries building at Wood's Holl (as it was then) to establish the Marine Biological Laboratory. The WEA also ensured that women might work at the laboratory by requiring that two of its members be on the board of trustees: the first two were graduates of Vassar and MIT.

Among Freeman's first appointments at Wellesley were Eliza Mosher as professor of practical physiology and Katharine Coman (also from Michigan) as professor of political economy and history. In 1885, she appointed Katharine Lee Bates an instructor of English. They were soon joined by Mary Calkins, a student of William James, who established the first laboratory of experimental psychology to be headed by a woman. Like Mosher, Coman, and Calkins, Katharine Lee Bates was destined to spend her entire life at Wellesley; she became full professor in 1891 and long-term chairman of the English Department until her retirement in 1925. But her life changed forever in 1887 when she met Katharine Coman. The two Katharines lived together for more than a quarter of a century in the loving bonds of what was then called a "Boston marriage" and is now appreciated as "a devoted lesbian couple." They called their Wellesley home the "Scarab," their faithful collie "Sigurd," and their automobile "Abraham" (because they were so often deep in its bosom). When parted by professional travel, they wrote passionate, almost daily, letters to each other:

> Your love is a proof of God. How does love come, unless Love is? . . . That is a glorious sentence wherewith to close your letter. I love it and I love you and I love what shadowy hint of God comes to me.

In 1893, on that journey to Pike's Peak, the two Katharines stopped to visit the great Columbian Exposition in Chicago, Bates becoming "naturally impressed by the symbolic beauty of the White City," as the grounds of the exposition were known. By that time Alice Freeman had become Alice Freeman Palmer after marrying a Harvard philosopher, and the Palmers together supervised the construction and installation of the Woman's Building, a monument fashioned in the mock alabaster of the exposition's Beaux Arts style. Featuring statues of prominent feminists such as Susan B. Anthony and Elizabeth Cady Stanton, its interior was chock-a-block with objects picked to show, as *The Book of the Fair* put it,

> . . . the contributions made by women to the huge workshop of which this world so largely consists, their contributions not only to the industries of the world but to its sciences and arts. Thus it is hoped in a measure to dispel the prejudices and misconceptions, to remove the vexatious restrictions and limitations which for centuries have held enthralled the sex.

The Woman's Building was an anomaly among the grander structures of the Columbian Exposition, the iconography of which played to the "prejudices and misconceptions" of centuries. In statuary great and small, burly men steered the ship of state, while women were placed on pedestals, perched as guiding spirits or cast as docile handmaidens. The Central Fountain, designed by Frederick MacMonnies, underscored each of these roles.

A visit to the Palmers' Woman's Building was not the only reason the two Katharines had stopped in Chicago; there were professional reasons as well. Bates had taught in Colorado Springs on equal terms with such male professors as William Rolfe of Harvard and Davis Todd of Amherst, and Coman had lectured on the economic history of Western expansion. In Chicago, Coman heard Frederick Jackson Turner deliver his famous address on "The Role of the Frontier in American History." Turner argued that the Western frontier had for three centuries been a metaphor for the American dream, for "manifest destiny." Now that the open frontier had closed and the United States had become one nation from sea to shining sea, other frontiers awaited. Coman's two-volume *Economic Beginnings of the Far West*

(1912) was devoted to parallel themes, and her view of the economic history of the railroads echoed Turner's message that the age of an external frontier and the West as wilderness was over. Coman and Bates were convinced in 1893 that the spirit that had won the West would in time remove those "vexatious restrictions and limitations."

THE AMERICA OF CONTINENTAL EXPANSION was no collection of White Cities. The year 1893 was a year of significant social unrest, and the strife was by no means liberating. Grover Cleveland entered the White House for the second time, with the country in the midst of a deep economic depression. On February 23, 1893, the Philadelphia and Reading Railroad had gone bankrupt, and before the end of the year the Erie, the Northern Pacific, the Union Pacific, and the Atchison, Topeka and Santa Fe went belly up as well. Two and a half million people—one fifth of the work force—were unemployed, and Henry Adams lamented that "much that had made life pleasant between 1870 and 1890 perished in the ruin." Even the president admitted that "values supposed to be fixed are fast becoming conjectural, and loss and failure have invaded every branch of business." In legislation that was to provide windfall profits for J. P. Morgan and Augustus Belmont, Cleveland brought the country back to the gold standard in the very week that Bates stood atop Pike's Peak.

Meanwhile, federal and state militias were sent against workers in the Homestead steel strike in Pennsylvania, against switchmen in Buffalo and coal miners in Tennessee, and finally against the Pullman strikers in Chicago. Each of these episodes of class warfare was later treated in Katharine Coman's *Industrial History of the United States* (1905).

The two Katharines were also in the vanguard of social activists. They were among the founders in 1887 of the College Settlements Association, a group that made it possible for young female college graduates to spend a year at community settlement houses among the poor and the immigrants—the "teeming refuse" of Europe's shores. In the course of this work Bates and Coman became closely associated with the pioneer of Chicago's Hull House—and future Nobel Peace Prize winner—Jane Addams. In 1889, Addams's lifelong companion, Ellen Gates Starr, described how the settlement-house movement might benefit not only the needy but also the philanthropist:

> It is not the Christian spirit to go among these people as if you were bringing them a great boon: one gets as much as one gives [but] people are coming to the conclusion that if anything is to be done towards tearing down these walls . . . between classes that are making anarchists and strikers the order of the day, it must be done by actual contact and done voluntarily from the top.

A generation of social workers, public-health activists, and egalitarians spent their lives convinced of the need for that actual contact.

In 1892, Katharine Coman became chairman of the committee that opened Denison House in Boston and made it a center of labor-organizing activity, to which Bates was inevitably drawn. Denison House, Hull House, and the other settlement houses were deeply committed to reform of working hours, protection of immigrants, compulsory school attendance, school health, and—above all—abolition of child labor. It was toward this end that the poet Sarah Cleghorn wrote in "The Masses":

> The golf links lie so near the mill,
> That almost every day,
> The laboring children can look out
> And see the men at play.

When violence broke out during the Chicago Pullman strike of 1894, and strikers burned down the remnants of the White City, Coman and Addams sided with workers against the militia; Coman went to Chicago again in 1910 to help striking seamstresses win union rights.

Meanwhile, tragedy had struck the couple; Coman's last work was completed as she lay dying of breast cancer. *Unemployment Insurance: A*

Summary of European Systems (1915) was a meticulous survey of how other industrialized countries cared for the aged, the disabled, and the unemployed. She concluded that social services in Bismarck's Germany and the Third Republic's France were far in advance of those in her own country. Posthumously published to little acclaim, Coman's book was to become a blueprint for social justice in the United States. Her call for old-age and disability benefits in the New World—social security—became a platform plank of the Progressive and then the Democratic party. After the New Deal, Coman's dream became the law of the land. Bates describes how she helped her friend finish this major economic study:

> Through those four years beset with wasting pain,
> The surgeon's knife again and yet again,
> . . . So we twain
> Finished your book beneath Death's very frown.
> For all the hospital punctilio,
> Through the drear night within your mind would grow
> Those sentences my morning pen would spring to meet . . .

These lines, and the ones following, come from a volume of passionate love poems, *Yellow Clover*, written by Bates and published in 1922, seven years after Coman died. Each of these poems was devoted to Katharine Coman, and in some Bates reached levels of emotional expression—perhaps even art—that eluded her in seven other volumes of verse.

> Your life was of my life the warp and woof
> Whereon most precious friendships, disciplines,
> Passions embroider rich designs . . .
>
> No more than memory, love's afterglow?
> Our quarter century of joy, can it
> Be all? The lilting hours like birds would flit
> By us, who loitered in the portico
> Of love's high palace . . .

Bates spoke in no loud voice the love that dared not speak its name: yellow clover stood for physical love in the flower language of the two Wellesley scholars, who

> Stooped for the blossoms closest to our feet
> And gave them as a token
> Each to each
> In lieu of speech,
> In lieu of words too grievous to be spoken . . .

"Undimm'd by human tears" is the hopeful lyric of Bates's most successful public poem, our national anthem of social justice, a hymn to the better angels of American nature. The lyrics of "America the Beautiful" should remind the bigots of this world of a generation of women whose emotional ties and social reforms have outlasted the alabaster cities' gleam of the Columbian Exposition.

> And crown thy good with brotherhood
> From sea to shining sea!

14.

Alice James and Rheumatic Gout

> . . . a lump I have had in one of my breasts for three months is a tumor. . . . This with a delicate embroidery of "the most distressing case of nervous hyperasthenia," added to spinal neurosis that has taken me off my legs for seven years, with attacks of rheumatic gout in my stomach for the last twenty, ought to satisfy the most inflated pathologic vanity.
>
> —Alice James (June 1, 1891)

IN NOVEMBER 1882, AFTER ONE OF MANY "NERVOUS" ATTACKS, Alice James (1850–1892) took to her bed in the Boston home of her father, Henry James, Sr. On the advice of her brother, Dr. William James, and after ten days of bed-ridden solitude, she agreed to consult one of her brother's colleagues at Harvard Medical School, Dr. Henry Harris Aubrey Beach. Dr. Beach suspected that there was "something lying in back of her nervousness," which coincided with her father's terminal illness. After three weeks of investigations, he informed his patient that her malady was "gout, rheumatic gout!"

The diagnosis was relatively new on this side of the Atlantic, having been put on the medical map by A. B. Garrod of London in 1859. Beach, however, was a very up-to-date physician; he had taught anatomy under Dr. Oliver Wendell Holmes, held an appointment at Massachusetts General

Hospital, and was an assistant editor of the *Boston Medical and Surgical Journal* (now the *New England Journal of Medicine*). Beach promised relief to his patient and, indeed, she was relieved to hear that her symptoms were due to something as physical as gout and not simply one more "fight between my body and my will." But in December her father died; she again relapsed into general debility, muscle weakness, and depression. Eventually she placed herself in the Adams Nervine Asylum, where for several months she was treated with electrical stimulators, vapors, and the rest cure of Dr. S. Weir Mitchell. These measures were of little avail and, after a variety of other treatments for symptoms attributed to "spinal neurosis," "nervous hyperaesthesia," "neurasthenia," or "suppressed gout," she crossed the ocean to join her brother Henry in England.

In London Alice encountered one of the few physicians who could explain to her that rheumatic gout had little to do with her lifelong infirmity. Dr. Beach had arranged the consultation, assuring the Jameses that Sir Alfred Baring Garrod (1819–1907) was the only man who knew anything about "suppressed gout." Beach was somewhat off the mark. Garrod, unquestionably the British authority on rheumatic complaints, thought that the term "suppressed gout" was gibberish. Today, Garrod is credited with turning the study of arthritis into the protoscience of rheumatology. At a public lecture on February 8, 1848, Garrod demonstrated that gout was due to a pile-up of uric acid in the blood and urine of the gouty, whereas there was no such increase in acute rheumatism (rheumatic fever). At the time, he was assistant physician at University College Hospital. Later, in 1854, he observed that linen strings dipped into the blood or urine of gouty patients became coated with visible crystals of uric acid. Garrod's "string sign" was a milestone in clinical biochemistry; similar deposits of crystals in the joints are the direct cause of gouty inflammation. Garrod went on to distinguish gout from rheumatic fever and, by the time he saw Alice James, he had just plucked a new diagnostic entity, rheumatoid arthritis, from the vague grab bag of conditions called "rheumatic gout." He wrote in his *Treatise on Gout*:

> Although unwilling to add to the number of names, I cannot help expressing a desire that one may be found for the disease under consideration, not implying any necessary relation to

> gout or rheumatism. . . . I propose the term rheumatoid arthritis, by which name I wish to imply an inflammatory affection of the joints, not unlike rheumatism in some of its characters, but differing materially from it in its pathology.

Alice James reported that she had "spent the most affable hour of my life" with Garrod, who told her in 1885 that the weakness in her legs and her digestive complaints were functional and not "organic" in origin. But—as usual with her doctors—she soon became disenchanted. At the dawn of the Freudian age she complained:

> I could get nothing out of him & he slipped thro' my cramped & clinging grasp as skillfully as if his physical conformation had been that of an eel instead of a Dutch cheese—The gout he looks upon as a small part of my trouble, "it being complicated with an excessive nervous sensibility" but I could get no suggestions of any sort as to climate, baths, or diet from him. The truth was he was entirely puzzled about me and had not the manliness to say so.

Whether like an eel or a Dutch cheese in manly conformation, Garrod was sensible enough not to suggest changes in "climate, baths, or diet." His patient suffered from no sort of gout known to him. Garrod was one of the first to draw clear distinctions between the common, heritable, and lead-induced varieties of gout. The lead-induced form, saturnine gout, had been rampant in the ruling classes of Europe since the sixteenth century and is caused by consumption of lead-contaminated spirits such as fortified wine or brandy. The disease took its greatest toll among the middle and upper classes, for the poor drank mainly gin and beer. Its epidemiology accounted for the notion of "the Honor of the Gout." Since saturnine gout was due to nurture, not nature, Garrod and his contemporaries had a good notion of what to do for other forms of rheumatic complaints such as "rheumatic gout": change nurture. Their well-off patients were packed off to take the cure at Bath or Leamington in England, or on the Continent, where a few weeks of lead-free water were expected to flush lead and uric acid out of their systems. Garrod so often prescribed hydrotherapy in Aix-les-Bains that today a street in the spa is named rue Sir Alfred Garrod. Garrod had seen enough patients like Alice James to conjecture that in the potpourri of conditions called rheumatic gout "there is much to show in its etiology and the distribution of the affected joints that is intimately connected with the nervous system."

Garrod and Alice James had a familiar sort of doctor–patient tussle over her symptoms. He believed that her muscle aches and cramps were caused by the galvanic stimulator she had used for several years; she believed that the Indian hemp (cannabis) prescribed by Garrod had made her sick. She sought refuge with a Dr. Townsend, who informed her that she had a "gouty diathesis complicated by an abnormally sensitive nervous organization." Finally, accompanied by her companion and a nurse, she took herself off to Royal Leamington Spa in 1889. She prepared to settle into the long, sessile routine of the invalid: *Alice in Bed* as legend (and the titles of two modern works) would have it. Clinging to the belief that while some of her symptoms might be those of rheumatic gout, she blamed the others on her own nervous temperament. "How well one had to be—to be ill!" she wrote in July 1890.

Nine months later, cancer struck, and by the spring of 1892, she was terminally ill:

> I am being ground slowly on the grim grindstone of physical pain and on two nights I had almost asked for K's lethal dose, but one steps hesitatingly along unaccustomed ways and endures from second to second.

The lethal dose would have been laudanum, and the K who dispensed it would have been Alice James's faithful companion, Katharine Peabody Loring. A few days after that entry, Alice James was dead from metastatic cancer, killed by what she called "this unholy granite substance in my breast."

She had known full well that this was something palpably different from the afflictions that brought her to Leamington:

> To him who waits, all things come! My aspirations have been eccentric, but I cannot complain now that they have not been

> brilliantly fulfilled. Ever since I have been ill, I have longed and longed for some palpable disease, no matter how conventionally dreadful a label it might have. . . . It is entirely indecent to catalogue one's self in this way, but I put it down in a scientific spirit to show that though I have no productive worth, I have a certain value as an indestructible quantity.

Indestructible she was not, but nevertheless she continued to produce in her journal an indestructible work of literature that ranks with those of her brothers. Discretion dictated that it appear eighteen years after Henry James's death (1934) after passing in the family to the daughter of Robertson, youngest and frailest of the James brothers. The journal is a record of daily events, an almanac that condenses a decade, and a sharp text of self-knowledge. It had begun simply enough: "I think that if I get into the habit of writing a bit about what happens, or rather doesn't happen, I may lose a little of the sense of loneliness and despair which abides by me."

But soon she found her own style, in which social insight, cutting wit, and a good nose for literature mixed well with a bent for reform and a tendency to what her brother Henry called "passionate radicalism." She was a sharp critic of conservative politics, a fan of Irish Home Rule and of Parnell. Hypocrisy outraged her:

> What a spectacle, the Anglo-Saxon races addressing remonstrances to the Czar against expelling the Jews from Russia, at the very moment when their own governments are making laws to forbid their immigration.

She was also ahead of her time in sniffing out the cost of empire. She worried about an English class structure in which:

> . . . the working man allows himself to be patted and legislated out of all independence; thus the profound ineradicables in the bone and the sinew conviction that outlying regions are their preserves, that they alone of human races massacre savages out of pure virtue. It would ill-become an American to reflect upon

> the treatment of aboriginal races; but I never heard it suggested that our hideous dealings with the Indians was brotherly love under the guise of pure cussedness.

After her tumor was diagnosed, the tone of her journal turned darker and the entries more confessional. She permitted herself at last to express openly her love for Katharine Loring. Loring was a competent, educated woman of Brahmin stock, of whom Alice James had written to Sara Darwin (Charles Darwin's American daughter-in-law),

> She has all the mere brute superiority which distinguishes man from woman combined with all the distinctively feminine virtues. There is nothing she cannot do from hewing wood and drawing water to driving run-away horses & educating all the women in North America.

Photographs show Katharine Loring as a tidy, angular woman: Alice B. Toklas as nurse and confidante. Loring had given up an active life to be with Alice James, shuttling between her own consumptive sister and the intermittently paralyzed Alice. In her last few months of life, the journal was no longer written by Alice James herself, but dictated to K. Alice James's tribute at what she called "this mortuary moment" is therefore perhaps the more poignant:

> . . . is it not wonderful that this unholy granite substance in my breast should be the soil propitious for the perfect flowering of Katharine's unexampled genius for friendship and devotion. The story of her watchfulness, patience, and untiring resource, cannot be told by my feeble pen, but the pain and discomfort seem a feeble price to pay for all the happiness and peace with which she fills my days.

Reading Alice's journal of disease and despair, a tableau comes to mind. It is set in a Kensington drawing room that might have been painted by John Singer Sargent, but the feeling is pure Edvard Munch. It is January 6,

1892, two months before her death, and Alice is racked by the jaundice of liver metastases. A photograph of the time shows Alice in her daybed; she is drawn and gaunt but has been dressed up and beribboned for the photographer. She sits stiffly propped on her cushions and appears pain-free for a while on morphine: the poppy is singing its song. She can no longer concentrate enough to write and dictates her journal entries to the beloved K. But K is by no means the only reader for whom these *pensées* are intended; she is also leaving a record for her brothers William and Henry.

Alice James contrasts the warm devotion of Katharine with the clinical cool of her doctors. Sir Andrew Clarke, who diagnosed that granite lump, is reprimanded for never quite being on time: he is of course the *late* Sir Andrew. But that eminent cancer specialist cannot help her; she is afraid that he is gripped by "impotent paralysis" and "talking by the hour without *saying* anything, while the longing, pallid victim stretches out a sickly tendril, hoping for some excrescence, a human wart, to catch on to . . . " Alice James utters a cry from what John Keats called a "world of pains and troubles" to the world of her loftier brothers:

> When will men pass from the illusion of the intellectual, limited to sapless reason, and bow to the intelligent, juicy with the succulent science of life?

Some day the science of life, biology, should be able to explain not only how tumor suppressor genes and hormones influence breast cancer, but also the lifelong affliction of souls as fragile as Alice James. The most convenient biological explanation is the genetic, and keepers of the Jamesian flame point to a striking cluster of nervous pathology in two generations of the family. Deep depression, hysterical paralyses, and sexual ambiguity seemed to run in the clan: father Henry suffered from a Swedenborgian "midnight vastation"; Henry, Jr., was accused by William of "coddled sensibilities" and "dorsal anguish"; William was paralyzed by "quivering fear." Mental homes provided refuge for the "incommunicable sadness" of the youngest brother, Rob, and the "madness" of cousin Kitty.

William James, the loftiest of the brood, believed that the family carried a genetic load on its back. Using the spinal imagery of his day, he wrote to

Rob, "I account it as a true crime against humanity for any one to run the probable risk of generating unhealthy offspring. For myself I have long since fully determined never to marry with any one . . . for this dorsal trouble is evidently s'thing in the blood." Alice also alluded to a family serpent motif, writing for publication:

> Dr. Tuckey [a mesmeric doctor, recommended by William] asked me the other day whether I had written for the press, I vehemently denied the imputation. How sad it is that the purely innocuous should always be supposed to have the trail of the family serpent upon them.

The family serpent was, of course, the double helix of depression and inspiration that ran through the James family, that "dorsal trouble . . . in the blood." The dying Alice James dictated to K an account of a visit paid her years before by Charles Darwin's daughter, Henrietta Litchfield. When Alice James told Mrs. Litchfield that her invalidism had for years been called "latent gout," Mrs. Litchfield exclaimed, "Oh! that's what we have [in our family], does it come from drink in your parents?" The Darwins, like the Jameses, were a clan of morose geniuses, scholars, and medics who seemed also to have had a family serpent in their blood. The Darwin biographies document a strong family history of depression, somatization reactions, and wives who took to their beds for decades; sure enough, much of this pathology was blamed on a "gouty diathesis." Depression and inspiration were closely intertwined in the Darwin line, as closely—one might say—as the two dancing serpents on the crest of Darwin College, Cambridge.

While social explanations rival the genetic, biography can describe, but not explain, the flight into sickness. Jean Strouse's authoritative biography was written before paired lives like those of Alice James and Katharine Loring were explored in the context of legal constraints and homophobia. Treating that subject by elision, Strouse argued plausibly that Alice James's flight into disease was at least in part a not uncommon female strategy for coping with oppressive male society in general and her father's expectations specifically. Strouse attributes some of Alice's symptoms to a daughter's reaction to a "kind father who had so blithely stimulated and thwarted her." She quotes Alice James's recollection of an acute early episode:

> As I used to sit immovable reading in the library with waves of violent inclination suddenly invading my muscles taking some one of their myriad forms such as throwing myself out of the window, or knocking off the head of the benignant pater as he sat with his silver locks, writing at his table, it used to seem to me that the only difference between me and the insane was that I had not only all the horrors and suffering of insanity but the duties of doctor, nurse and strait-jacket imposed upon me, too.

In retrospect, no one can blame with confidence either nature or nurture for Alice James's life of infirmity and her unasked-for crucifixion, to use a phrase of Oliver Sacks's. Nurture in the form of nineteenth-century medical practice was clearly responsible for the diagnoses pinned on Alice James in her lifetime: spinal irritation, neurasthenia, hysteria, suppressed gout, and so on. Medical practice also subjected her to electric prods, Indian hemp, spinal manipulation, and all those buckets of tepid spa water . . . "as if to be singed and scalded were a costly privilege and leeches were a luxury," a phrase of Dr. Oliver Wendell Holmes.

Alice was one of many poor spirits who were flogged from pillar to post on the premise that their paralyses and palpitations were due to rebel humors in the spine; the tale of their prodding, buzzing, and poking has been told by the historian of our psychosomatic era, Edward Shorter. Those treatments were not simply an assault of male doctors on their spinally challenged female patients. The male version of hysteria was called hypochondriasis, and the paralyses it produced were also, willy-nilly, attributed to spinal irritation. Fresh from Harvard Medical School, Dr. William James advised Rob, the most nervous of the James brothers, to take iron, to exercise, and to apply iodine to his back until his skin peeled. The doctrine of counterirritation demanded the show of wounded flesh on the part of male and female patient alike. That doctrine remains a major feature of folk medicine the world over; Western medicine abandoned it when we gave up cupping.

In the end, her mother, Mary, knew what was wrong with Alice James: "It is a case of genuine hysteria for which no cause as yet can be discovered." Alice herself awaited a new science of the mind. In her last letter to William, whose *Principles of Psychology* had just been published, she pleaded, " . . . so when I'm gone, pray don't think of me simply as a creature who might have been something else, had neurotic science been born."

These days, we may be better at the diagnosis and treatment of breast cancer, and the laws of Massachusetts have changed with respect to paired lives, but we are not much further along in neurotic science. From patients humbler than Alice James, we've learned that one does not need a "palpable disease" to be in real pain. Patients with psychosomatic diseases are often in anguish, and good doctors address the anguish and not the diagnosis. My own guess is that nowadays Alice James's illness would have been diagnosed as fibromyalgia, chronic fatigue syndrome, irritable bowel syndrome, or another one of man's "medically unexplained diseases." The British psychiatrist Simon Wessely of King's College London explains that most medical specialists define unexplained syndromes in the technical terms of their own expertise. Presented with the same cluster of symptoms, what a rheumatologist will call "fibromyalgia" a gastroenterologist would diagnose as "irritable bowel syndrome," while a neurologist might come up with "chronic fatigue syndrome" and an infectious disease specialist might test for "chronic Lyme disease." Wessely provides convincing evidence that none of these monikers describes a unique clinical entity; each syndrome shares much with all the others. From epidemiological evidence he concludes that there are strong associations between persistent symptoms such as muscle weakness, stomach cramps, and overall fatigue—the symptoms of Alice's "rheumatic gout," one might say. A safer bet, he argues is to describe these conditions honestly as "medically unexplained syndromes."

Unexplained or not, one more set of medical diagnoses intrudes on the story. A. B. Garrod left us a legacy as imposing as his work in the rheumatic diseases. His son, Sir Archibald Edward Garrod (1857–1936), followed his father's teachings on the heritable nature of gout. He became interested in a series of rare genetic disorders that, in his Croonian Lectures, he called "inborn errors of metabolism." And it is thanks to modern studies of inborn errors of metabolism—phenylketonuria is a prime example—that the principle emerged of "one gene, one enzyme." By means of that principle, Arthur Pardee, François Jacob, and Jacques Monod were able in 1959 to discover how changes in nutrients outside a bacterium can switch single, transmissible genes on or off within a cell: an example of how nurture modifies nature in the dish. This was one of the milestones of molecular biology.

Nowadays, when we treat patients with phenylketonuria by means of an appropriate diet from birth, we can prevent both physical and mental disease, a striking example of how nurture can modify nature in the clinic as well. Clues from the study of depression suggest that the family serpent of gloom in the blood may also turn out to be an inborn error of metabolism. The irony would not have escaped either William or Alice James that the Garrods, who began the science of gout, were present at the birth of "neurotic science."

A last irony in the story of Alice James is that her legacy is one she could not have imagined. Far better known today than in her lifetime, she has been variously celebrated as a model memoirist, a feminist critic of Victorian mores, a pioneer of narrative medicine, and a lesbian heroine. Over the years, she has become an exemplar in arguments over the role that male physicians play in imposing diseases on women, the role of gender in renown, and the requirement for victimhood in literary honor. Her journal fits into a genre Elaine Showalter has called *Hysteries* (1998). Cathleen Schine's first novel, *Alice in Bed* (1983), is about a woman hospitalized for a medically unexplained disease; the doctors are demons. Susan Sontag's more recent play, also titled *Alice in Bed* (1993), is a more fitting tribute; it begins like Gertrude Stein and ends in a benediction. The play opens with the voice of Alice's nurse in Leamington:

> NURSE: Of course you can get up
> ALICE: I can't
> NURSE: Won't
> ALICE: Can't
> NURSE: Won't

Later Alice soliloquizes about Rome, a city that she, unlike her famous brothers, has never seen: "I would be very humble. Who am I, compared with Rome. I come to see Rome, it doesn't come to see me. It can't move." The play ends with Alice saying, "Let me fall asleep. Let me wake up. Let me fall asleep." To which her nurse replies, "You will."

She wakes up on every page of her journal.

15.

Free Radicals Can Kill You: Lavoisier and the Oxygen Revolution

Mme Lavoisier: Imagine what it means to understand what gives a leaf its color! What makes a flame burn. Imagine!

—Carl Djerassi and Roald Hoffmann, *Oxygen* (2001)

Oxygen is nothing other than the most salubrious and purest portion of the air, such that . . . it appears in an eminently respirable state more capable than the air of the atmosphere of sustaining ignition and combustion.

—Antoine-Laurent Lavoisier (1775)

. . . though pure dephlogisticated air [oxygen] might be useful as a medicine, it might not be so proper for us in the usually healthy state of the body; for as a candle burns out much faster in dephlogisticated than in common air, so we might, as may be said, live out too fast, and the animal powers be too soon exhausted in this pure kind of air. A moralist, at least, may say, that the air which nature has provided for us is as good as we deserve.

—Joseph Priestley (1775)

> . . . there are good reasons for assuming that the changes produced by irradiation and those which arise spontaneously in the living cell have a common source—the OH and HO_2 radicals.
>
> —Denham Harman (1956)

I'M GETTING WORRIED ABOUT OXYGEN: not having too little of it, but too much. We've known since Lavoisier that flames burn, metals rust, and we take breath—all thanks to oxygen. But we hadn't learned how oxygen excess does harm until 1954, when Rebecca Gerschman et al. worked out that oxygen poisoning and x-irradiation share the property of producing oxygen-derived free radicals. Add one electron to oxygen, as by irradiation, and superoxide anion ($O_2^{-\bullet}$) is formed; that's a free radical anion. Add two electrons, and you get a bleach, hydrogen peroxide (H_2O_2), which is not a free radical but can readily react with $O_2^{-\bullet}$ to form the very reactive hydroxyl radical ($OH^\bullet$). Add any of these to living tissues, and they do damage. On the basis of this chemistry, Denham Harman formulated his heuristic "free radical theory of aging":

> . . . In regard to aging, I felt that there had to be a basic cause which killed everything. Further, this basic cause should be subject to genetic and environmental influences.

It is no accident that Gerschman's and Harman's laboratories were funded by the Atomic Energy Commission. As Harman later observed, the work "was of particular interest in 1954 because of concern over possible nuclear war." Well, thanks to the Cold War and the last fifty-odd years of research in the area, we've learned that Priestley was right about that burning candle. We've learned that we do "live out too fast," and our animal powers are indeed "too soon exhausted" when we are bombarded by oxygen-derived free radicals, whether generated from without or within our own bodies.

The flood of publications dealing with oxidative stress has risen dramatically in recent years. Scanning papers with "oxidative stress" in the title, I checked out the website Web of Science. Sure enough, between 2005 and mid 2017, papers dealing with oxidative stress had increased 300-fold. In 2005, papers dealing with low oxygen (hypoxia) exceeded in popularity the hyperoxic condition, but by mid-2017, articles on oxidative stress (279,243) were beating hypoxia (122,644) by more than 2:1. Priestly would have been pleased.

MANY OF THOSE PAPERS SPELL OUT THE FACT that we age by the same mechanism by which metals rust, photos fade, and wicker frays. They also confirm Harman's prediction that in our bodies, these processes are "subject to genetic and environmental influences." We know that genetic defects in one or another of our own defenses against those free radicals, such as superoxide dismutase or ceruloplasmin, cause serious disease. Our cells also make reactive oxygen species, not only in the course of mitochondrial respiration within cells, as already surmised by Gerschman et al. Our inflammatory cells also assemble machinery for making $0_2^{-\bullet}$ and $OH^\bullet$ on the cell surface. In concert with a newly recognized brew of reactants formed from NO, ozone, and so on, these wreak havoc with our cells and extracellular constituents. Lavoisier had the chemistry right when he told us what happened to the animal "oils" of our cells when oxygen reacts:

> It is evident that the oils, being composed of hydrogen and charcoal combined, are true carbono-hydrous or hydro-carbonous radicals; and, indeed by adding oxygen, they are convertible to vegetable oxides and acids according to their degrees of oxidation.

And that was before hydroperoxy and peroxy fatty acids—our time's version of vegetable oxides—were on our screens.

ANTOINE-LAURENT LAVOISIER AND HIS WIFE, Marie-Anne, are depicted in the most beautiful image of scientific coworkers ever put on canvas. Jacques-Louis David's 1788 portrait of this unusual couple is not only an image of Enlightenment grace and grandeur but also a document of a work in progress for which both of the principals are responsible—the revolutionary *Traité élémentaire de chimie* of 1789. It justifies the later appellations of Lavoisier as the father and Mme Lavoisier the mother of modern chemistry.

Lavoisier is shown with quill on paper, looking up at his wife as if to take dictation or suggestion. The instruments on the table and floor are those used by Lavoisier to give "the first accurate accounts of burning, respiration and rusting." Mme Lavoisier is depicted with her arm on her husband's shoulder, and the stand behind her could well contain drawings for some of the thirteen plates she fashioned for the *Traité de chimie*. Lavoisier's mid-script attention to his wife reminds me of her important translation of Richard Kirwan's treatise on the presumed substance "phlogiston". Her modest smile directed at the viewer—and the portraitist—suggests a strong student–teacher bond. Marie-Anne was an apt pupil of David: her skillful student drawings, with David's comments, are in the Musée des Arts et Métiers in Paris today.

The painting also documents a concurrence of three revolutions: the American, the French, and the Chemical. It is set in the Arsenal of Paris, to which Lavoisier had been appointed as commissioner of the Royal Gunpowder and Saltpeter Administration. His charge, so to speak, was to greatly improve the purity and efficacy of French explosives. He fulfilled this task admirably: in aid of his American friends, Thomas Jefferson and Benjamin Franklin, he guaranteed the colonists a trusty supply of neat gunpowder to fight the redcoats. Indeed, Franklin was a good friend of both Lavoisiers. Recovering from an attack of gout in Philadelphia, Franklin wrote to Marie-Anne in 1783 to thank her for a portrait she had painted of him for its "great merit as a picture in every respect; but what particularly endears it to me, is the hand who drew it." He was only one of Marie-Anne's many admirers, who included Gouverneur Morris, Pierre Samuel du Pont de Nemours, and Benjamin Thompson, Count Rumford. As Roald Hoffmann lamented, "There is no biography of Mme Lavosier. I think she deserves an opera."

Dispute has raged over which of the three chemists who came across oxygen between 1772 and 1778 deserves credit for the air of life. It is now agreed that a Swede discovered it first, the "fire air" of Carl Wilhelm Scheele in 1772; a Briton published it first, the "dephlogisticated air" of Joseph Priestley in 1775; and a Frenchman understood it first, the "oxygen" of Lavoisier in 1775–1778. J. W. Severinghous suggests that Lavoisier may have heard about Scheele's and/or Priestley's work directly or indirectly. But there's surely credit enough for all three to split the "retro-Nobel" awarded by Carl Djerassi and Roald Hoffmann in their sparkling play, *Oxygen*

At the Arsenal, Lavoisier extended his work on "oxygen" (he gave it the name from the Greek for "acid-forming") as the mediator of fire and life. He also proposed the first systematic enumeration of elements, a precursor of Mendeleev's periodic table:

> Thus, while I thought myself employed only in forming a Nomenclature, and while I proposed to myself nothing more than to improve the chemical language, my work transformed itself by degrees, without my being able to prevent it, into a treatise upon the Elements of Chemistry.

The treatise helped him to formulate the law of the conservation of matter: nothing is lost in a chemical reaction. In perhaps his most vital work, he showed that living beings transform oxygen in the course of respiration, that they consume energy and generate heat, and that muscular exercise burns calories as a candle does. One can measure this process. He called it calorimetry.

Alas, Lavoisier fell to the guillotine, not for his science but for his business interests. Lavoisier was one of approximately two dozen partners—among them, Mme Lavoisier's father—in a private, for-profit corporation, the Ferme générale. The Ferme functioned as a crop-inspection and tax-collecting agency working on behalf of the crown. The proceeds from Lavoisier's share of Ferme générale revenue are said to have paid for his experiments at the Arsenal.

After July 14, 1789, Lavoisier became a staunch supporter of the liberal constitutional monarchy set up in response to the fall of the Bastille. In the upbeat interregnum of 1790, he wrote to Benjamin Franklin of the two revolutions that were still in progress. Lavoisier, supported by his colleagues at the Academy of Sciences, had overthrown the reigning "phlogiston" theory. Phlogiston had been tautologically defined as a "fire-like" element in any substance undergoing combustion, but Lavoisier showed that the fire-enabling substance could come from air: oxygen. Lavoisier exulted to Franklin: "Here then a revolution has taken place in an important part of human knowledge since your departure from Europe." He was equally happy with the change in French political life: "After having brought you up to date on what is going on in chemistry, it would be well to speak to you about our political revolution. We regard it as done and without possibility of return to the old order."

THE OLD ORDER DIDN'T RETURN, but the Reign of Terror descended, and the tumbrels rolled. By 1792, the radical Jacobins were calling for heads, not only of the body politic. In quieter times, Jean-Paul Marat, that self-proclaimed genius of "optics, soap bubbles and medical electricity," had been denied admission repeatedly to the Academy of Sciences. Unleashed by the Terror, he called for abolition of all of the academies. In his broadsheet, *L'ami du peuple*, he attacked Lavoisier personally:

> I denounce . . . the leader of the chorus of charlatans, Sieur Lavoisier, son of a land-grabber, apprentice-chemist, pupil of the Genevan stock-jobber [Jacques Necker, Louis XVI's finance minister], a Fermier-général, Commissioner for Gunpowder and Saltpeter, Governor of the Discount Bank, Secretary to the King, Member of the Academy of Sciences.

Together with the constitutional monarchy, the Ferme générale was swept aside in the course of the Revolution. The radical Jacobins moved from mass protest to mass murder in what came to be called the "September massacres" of 1792. Lavoisier's laboratory at the Arsenal was shut forever, and the Academy of Sciences dissolved shortly thereafter. In November 1793, Lavoisier, his father-in-law, and twenty-six others of the Ferme générale were imprisoned and accused of "having plundered the people and the treasury of France, and of having adulterated the nation's tobacco with water, etc." All were found guilty after a one-day trial and condemned to the guillotine on May 8, 1794.

Two apocryphal quotations survive from the trial. One is attributed to Jean-Baptiste Coffinhal du Bail, the judge who dismissed appeals that cited Lavoisier's scholarly contributions to chemistry and the nation: "*La République n'a pas besoin de savants ni de chimistes!* (The Republic needs neither scholars nor chemists!)" The other, attributed to the mathematician Joseph-Louis Lagrange, may be more authentic:

> It took them only a moment to sever that head, and a hundred years perhaps will not suffice to produce another like it.

Close enough: Albert Einstein was born in 1879.

16.
Dr. Blackwell Returns from London

> You ask, what use will she make of her liberty when she has so long been sustained and restrained? I answer in the first place this will not be suddenly given. . . . But were this freedom to come suddenly, I have no fear of the consequences. . . . If you ask me what offices they may fill, I reply—any. I do not care what case you put; let them be sea captains if they will. I do not doubt there are women fitted for such an office. . . .
>
> —Margaret Fuller, *Woman in the Nineteenth Century* (1845)

MARGARET FULLER AND HER HUSBAND AND INFANT SON were drowned by shipwreck on July 19, 1850, when an inexperienced sea captain ran aground the ship in which they were traveling on a sandbar off Fire Island, New York. Exactly one year later, another American woman returned to America from postgraduate medical studies in London to a reception only somewhat more hospitable than Fuller's. Dr. Elizabeth Blackwell (1821–1910), described by the *Lancet* as "the first woman medical graduate in the modern meaning of the phrase," arrived in New York to set up medical practice. She, too, followed her medical vocation at the École de Médecine, where Dr. Holmes had studied, while her life in reform realized

George Eliot's hope that the profession would combine the goals of intellectual conquest with social good. In several aspects Dr. Blackwell's career paralleled that of Dr. Holmes. Children of preachers, both had imbibed the sternest of Puritan values, both softened their views in the light of French culture, both lived in the service of public hygiene. Both strongly opposed the heresies of Mesmer and homeopathy, and both strongly believed that meliorist reason would "increase the power of positive good" in the new republic.

Born in England to Samuel Blackwell, a well-off sugar refiner and dissident lay preacher, Elizabeth and her eight siblings were brought to live in Cincinnati, where family friends soon included Henry Ward Beecher and Harriet Beecher Stowe. Her exposure to Unitarian thought in Cincinnati and to Quaker physicians in Philadelphia turned her interests to medicine, and she received medical tutorials in the private practices of friendly doctors. Despite thorough preparation in anatomy classes and a solid educational record, she was refused admission by every medical faculty in Philadelphia, New York City, and Boston, and by Bowdoin and Yale.

However, there was a small medical faculty in Geneva, New York, that was empowered to give the degree of doctor of medicine if a candidate attended lectures for two years and wrote a thesis. The school's requirements for the MD degree were par for the course, and the faculty put the question of a woman's admission to the students of the little upstate school; to their credit, the young men passed the following two resolutions, a copy of which remained with Blackwell to her death:

> 1. *Resolved:* That one of the radical principles of a Republican Government is the universal education of both sexes; that to every branch of scientific education the doors should be open equally to all; that the application of Elizabeth Blackwell to become a member of our class meets our entire approbation; and in extending our unanimous invitation we pledge ourselves that no conduct of ours shall cause her to regret her attendance at this institution.

> 2. *Resolved:* That a copy of these proceedings be signed by the chairman and transmitted to Elizabeth Blackwell.

Blackwell matriculated in November 1846, was more or less well received by town and gown, and performed splendidly in all her courses, especially therapeutics. In the summer of 1848, she took her clinical instruction at the Philadelphia Hospital, where an epidemic of typhus was in progress. This outbreak in an Irish immigrant population prompted her to collect the records of its victims and to describe how it was spread from case to case; the account became her doctoral thesis. Recommending light, air, and soap, her small treatise is only somewhat less professional than Holmes's on puerperal fever. Blackwell's thesis also relied on the work of Dr. Charles Louis, quoting his 1829 book on the distinction between typhus and typhoid. (That is the volume, we recall from *Middlemarch,* to which Lydgate turns before he gives up his dreams of research to marry Rosamond Vincy.) By February 1849, Blackwell's thesis had been published in the *Buffalo Medical Journal and Review,* and all that remained for her doctorate was to finish the two-year course of study, which she did with distinction. Her younger brother Henry described the scene of her graduation:

> The President taking off his hat rose, and addressing her in the same formula [as the others but] substituting *Domina* for *Domine,* presented her the diploma, whereupon our Sis, who had walked up and stood before him with much dignity, bowed and half turned to retire, but suddenly turning back replied: "Sir, I thank you; by the help of the Most High it shall be the effort of my life to shed honor upon your diploma."

The occasion was an event in both the United States and England, and the press by and large commented favorably. London's *Punch* chimed in with lines that would have made Rosamond Vincy cringe:

> Young ladies all, of every clime
>   Especially in Britain,

Who wholly occupy your time
In novels or in knitting,
Whose highest skill is but to play
Sing, dance, or French to clack well,
Reflect on the example, pray,
Of excellent Miss Blackwell! . . .

For Doctrix Blackwell—that's the way
To dub in rightful gender—
In her profession, ever may
Prosperity attend her!
Punch a gold-handled parasol
Suggests for presentation
To one so well deserving all
Esteem and admiration.

The degree won, Blackwell determined to become a surgeon. She was advised to seek the best clinical training possible and told that France was the place to obtain it. As we have seen, young American doctors properly regarded Paris as the fount of clinical science, and plucky Blackwell sailed off to the City of Light. Her adventures in Paris and afterward are described in a sparkling memoir, *Pioneer Work in Opening the Medical Profession to Women,* a volume that can stand comfortably on the shelf of meliorist literature somewhere between Thomas Wentworth Higginson's *Army Life in a Black Regiment* and *Middlemarch.* Blackwell picked Paris as

> the one place where I should be able to find unlimited opportunities for study in any branch of the medical art. . . . On May 21, 1849, with a very slender purse and few introductions of any value, I found myself in the unknown world of Paris, bent upon the one object of pursuing my studies, with no idea of the fierce political passions then smoldering amongst the people, nor with any fear of the cholera which was then threatening an epidemic.

Like young Oliver Wendell Holmes, who had come to Paris after the July Revolution of 1830, she arrived in the middle of social unrest. Like Holmes, she was thrilled by French sights and sounds; unlike Holmes, her first purchase was a new bonnet, "choosing plain grey silk, although I was assured again and again that nobody wore that color." Blackwell required a new bonnet because, again unlike Holmes, she was no unknown quantity; and her few introductions of any value included one to the poet Alphonse Lamartine, the short-term head of the short-term Second Republic. Tocqueville had said of him, "I do not think I ever met in the world of ambitious egoists in which I lived any mind so untroubled by thought of the common good as his." Blackwell was only somewhat more impressed:

> I was asked if I was a lady from America, for Lamartine is to most people *in the country.* I was shown through several antechambers into a drawing-room, where stood the poet entertaining some visitors, he bowed, requested me to wait a few moments and withdrew to his apartment: a lofty room, carved and richly gilded, three long windows opening on to a balcony and commanding a garden full of trees. The room contained a rich carpet and purple velvet couches, some portraits, an exquisite female profile in bas-relief, a golden chandelier from the ceiling, some antique vases, etc. and a soft green light from the trees of the large garden suffused the room.

Lamartine received her with Gallic poise, appearing "very tall and slender, but the most graceful man I have ever seen, every movement was like music; grey eyes and hair." Blackwell transmitted messages to him from friends in Philadelphia of the French Republic; Lamartine made amiable chatter in English—his wife was English—and said he was pleased to have these expressions of solidarity from friends of reform overseas. "There was perfect harmony in the man and his surroundings. Doubtless he is a true man, though unable to work into practice the great thoughts he cherishes." So went the meeting between the liberal poet-politician of France and the first Anglo-Saxon female physician. (Her book makes livelier reading these days than his poems.)

The next letter brought to her door none other than Dr. Louis himself, "then at the height of his reputation." She felt instinctively that his visit was one of inspection. She told him she was in need of practical work in surgery, and after a long conversation he informed her of what she must do to be permitted to work at La Maternité hospital, where she would in a very short time become expert in "one small field of surgery, obstetrics and gynecology." Shortly thereafter, with Louis's intervention, she was admitted to the Maternité in the autumn of 1849 for a six-month course.

In the enclosed world of the Maternité, teaching and practice were conducted by midwives (*sages-femmes*) in a convent-like atmosphere supervised by men. Her most intellectually stimulating companion was neither a midwife nor the professor in charge but the intern Hippolyte Blot. The two exchanged lessons in English and histology, spending hours at the young man's microscope. Blackwell wrote in *Pioneer Work,*

> By such examination different formations can be distinguished from each other; thus cancer possesses very distinctive elements. It is necessary to examine bodies of varying shapes under different foci of the microscope, otherwise illusions may be created. In illustration he placed some blood globules and showed us that what appeared to be a central spot in each globule was owing to the convexity not being in focus, and it disappeared when the focus was a little lengthened.
>
> He is busy himself now in preparing for an examination of *internes*; if he gains the gold medal, he has the right to enter any hospital he chooses as an *interne* for a second term, and receive also his M.D., not otherwise granted to an *interne.* What chance have women shut out from these instructions? Work on, Elizabeth!
>
> Today M. Blot spoke of a friend, Claude Bernard, a distinguished young inquirer, who is now, he thinks, on the eve of a discovery that will immortalize him . . . of the power which the liver has of secreting sugar in a normal state when animals are fed on certain substances which can be so converted; also of the

> curious experiment by which a dog was made, in his presence, to secrete albuminous or diabetic urine.

Claude Bernard (1813–1878), who may be called the founder of experimental medicine, once remarked that science is like a brilliantly lighted banquet hall which can only be reached after walking through a warren of ghastly and ill-lit kitchens. Young Drs. Blot and Blackwell were in no bright hall at the Maternité. They were like Rosencrantz and Guildenstern in a Denmark of physiological chemistry, and Bernard's finding that blood sugar came from liver glycogen was a towering Elsinore of mid-century science. This discovery, now called glycogenolysis, has led to an understanding of sugar and energy metabolism as well as of diabetes, did in fact make the "young inquirer" immortal, and one can construct a line of descent from Claude Bernard to the heroes of modern molecular biology—to James Watson and Francis Crick, François Jacob and Jacques Monod, David Baltimore and Paul Berg and Jim Bishop and Harold Varmus.

With her appetite for the new, young Dr. Blackwell in Paris tasted not only real science but also the mock. On a rare day off from the *sages-femmes,* she visited her sister Anna, who was living with another young American woman on the rue Fleurus (a street destined to become popular with American women: it was the future home of Gertrude Stein and Alice B. Toklas). From there the sisters went to a magnetic séance at the atelier of a socialite mesmerist, the Baron Dupotet. He, too, was at the height of his power, having recently converted half the alienists of England to the cause of what was to be known as "hypnosis."

The Blackwells were brought to a large, darkened room at the top floor of a large house on the Île de la Cité. A great portrait of Mesmer himself dominated the antechamber of the baron's quarters, its frame encrusted with firebrands and anchors and other significant images; he looked fixedly at a pale lady opposite to him who had evidently undergone several magnetic crises. A great number of indecipherable verses were tacked to the walls and hanging from the ceiling. In this large waiting room, a shifting population of fifty or so people went to and fro. This included a lady with a hole in her cheek, the painter to the king of Sweden, the son of the English consul in Sicily, and "a remarkable fat dame, seated just within the folding-doors, who had powerful fits of nervous twitching, which gave her a singular appearance of pale tremulous red jelly," as Elizabeth recalled in *Pioneer Work.* The baron's inner chamber was smaller and ornamented by mystic symbols and black letter books of the Black Art. There was housed the inevitable oval metallic mirror "traced with magic characters which exerts a truly wonderful effect upon impressionable subjects, exciting an ecstasy of delight or a transport of rage." One or another of the crowd would rush to grab it from the mesmerist; small amiable spats broke out over its possession.

No miracles were wrought that day, Blackwell assures us. Nevertheless, the faithful audience hung with great interest on every example of hypnotism: "the aspiring features assumed a higher aspect, the downward ones bent more determinedly, and the red jelly became more tremulous at every fresh magnetization; and when the *séance* closed everybody shook everybody's hand, and found it good to have been there." Blackwell judged the baron an honest man who for twenty-five years had been pursuing difficult and arcane subjects; "how much truth he may possess I am quite unable to say, for my position . . . has given me no power of really investigating them."

Back with her patients at the Maternité for only a few days, a grave accident befell Dr. Blackwell: "in the dark early morning, whilst syringing the eye of one of my tiny patients for purulent ophthalmia," some of the liquid spurted into her left eye. By nightfall on November 4, the eye had become swollen, and by the next morning, the lids were "closely adherent from suppuration." The diagnosis of purulent ophthalmia, the dreaded venereal disease of newborns and those who attended them, was made by a young staff physician, and the twenty-eight-year-old Blackwell was placed in the student infirmary.

We now know that the disease is caused by the gonococcus bacterium, is due to chronic gonorrheal infection of the female reproductive tract, and was part of the load borne by the prostitutes and working women who gave birth in the public hospitals of Paris. The bacteriological revolution has all but eliminated it, but Albert Neisser did not discover the microbe until 1879, and it was not until 1884 that Karl Credé showed that eyedrops of 1% silver nitrate on the lids of newborns were an effective prophylactic. Thanks to rigorous maternal health laws, by the beginning of the twentieth century prophylaxis had pretty much eliminated neonatal ophthalmia from advanced countries; today, unfortunately, it is making a comeback in Africa and Asia. But this was 1849, and Elizabeth Blackwell was treated by accepted methods of the day: cauterization of the lids, leeches to the temple, cold compresses, ointment of belladonna, opium to the forehead, purgatives, and footbaths. She was placed on a broth diet, and the eye was syringed every hour. "I realized the danger of the disease from the weapons employed against it," she remembered. Her friend Blot consulted his chief and he was given permission to devote the first days of the illness entirely to her case. He came in every two hours, day and night, to tend the eye. But despite his efforts, after three days it became obvious to her doctors that the eye was hopelessly infected. "Ah! how dreadful it was to find the daylight gradually fading as my kind doctor bent over me and removed with an exquisite delicacy of touch the films that had formed over the pupil! I could see him for a moment clearly, but the sight soon vanished, and the eye was left in darkness."

She lay in bed with both eyes closed for three weeks, but then the right eye gradually began to open and she could start to do little things for herself. She immediately wrote to her uncle, an English army officer—her father had died in Cincinnati when she was much younger—assuring him of her resolution to continue her career in medicine: "I beg uncle to feel quite sure that a brave soldier's niece will never disgrace the colours she fights under but will be proud of the wounds gained in a great cause." She downplayed her injury and told him that the accident could have been worse—the left eye was not greatly disfigured and would in time appear less so. She finished her letter with the assurance that she could write without difficulty, read a little, and hoped very soon to return to her studies: "I certainly esteem myself very fortunate and still mean to be at no very distant day *the first lady surgeon in the world.*"

As soon as she was up and about, she conspired with her sister to find a present for Hippolyte Blot, whose constant attention and compassion touched her deeply. "My friendship deepened for my young physician, and I planned a little present for his office." A very elegant pair of lamps were secured by Anna, which Elizabeth, she wrote, "bundled up in my dressing gown and shawl, looking and feeling very much like a ghost," hurried through the corridors to receive. That night she brought the lamps to Blot's consulting room, and in the morning, the young intern "came to me evidently full of delight, and longing to be amiable, yet too conscientious to infringe the rules of the Maternité by acknowledging the present." She was discharged on November 26 permanently blind in one eye. Despite her passionate ambition to be "the first lady surgeon in the world," because of this handicap she disqualified herself from surgery or obstetrics as a career.

She went to various spas in Germany to convalesce, and while there resolved to continue her training in medicine. She applied to St. Bartholomew's Hospital in London, then perhaps the strongest teaching hospital of the city. The illustrious English physician Sir James Paget endorsed her admission as a student "in the wards and other departments of the hospital," and on May 14 she was accepted at Bart's. Once on the wards, her zeal for medical science was as high and her dreams of reform as ambitious as those of any Lydgate or Holmes. She soon spotted the difference between the medicine of Paris and London at mid-century:

> I do not find so active a spirit of investigation in the English professors as in the French. In Paris this spirit pervaded young and old, and gave a wonderful fascination to the study of medicine, which even I, standing on the threshold, strongly felt. There are innumerable medical societies there, and some of the members are always *on the eve* of most important discoveries; a brilliant theory is *almost* proved, and creates intense interest; some new plan of treatment is always exciting attention in the hospitals, and its discussion is widely spread by the immense crowds of students freely admitted.

Bart's had its pleasures as well. Blackwell was courteously treated, saw the best of empirical medicine, and walked with men who, while not experimental or quantitative in approach, still felt clinical medicine in their sinews and knew how to examine patients. She wrote to her sister:

> This famous old hospital is only five minutes' walk from my lodgings, and every morning as the clock strikes nine, I walk down Holborn Hill, make a short cut through the once famous

> Cock Lane, and find myself at a gate of the hospital that enables me to enter with only a side glance at Smithfield Cattle Market. . . . Mr. Paget spoke to the students before I joined the class. When I entered and bowed, I received a round of applause. My seat is always reserved for me and I have no trouble. There are so many physicians and surgeons, so many wards, and all so exceedingly busy, that I have not yet got the run of the place; but the medical wards are thrown open unreservedly to me either to follow the physician's visits or for private study.

She also saw some of the same mock science that had flourished in France: mesmerism, homeopathy, and hydropathy, which she called the three heresies to distinguish them from the old system. But, after looking into the heresies a little more closely, she felt as dissatisfied with them as with what she had been taught: "We hear of such wonderful cures constantly being wrought by this and the other thing, that we forget on how small a number the novelty has been exercised, and the failures are never mentioned; but on the same principle, I am convinced that if the old system were the heresy, and the heresy the established custom, we should hear the same wonders related of the drugs." There is more than an echo here of Holmes's "If all the Materia Medica as now used, could be sunk to the bottom,—it would be the better for mankind and all the worse for the fishes." Sound advice in the days before penicillin.

Her mentor gave her some advice, which elicited a passionate response:

> Mr. Paget who is very cordial, tells me that I shall have to encounter much more prejudice from ladies than from gentlemen in my course. I am prepared for this. Prejudice is more violent the blinder it is, and I think Englishwomen seem wonderfully shut up in their habitual views. But a work of the ages cannot be hindered by individual feeling. A hundred years hence women will not be what they are now.

Blackwell's experiences in Paris and London made her eager to start out on her own in America. She wrote of her plans to her sister Emily in

November 1850 (Emily had decided to follow in her sister's footsteps and was being privately tutored by Dr. John Davis, an anatomy demonstrator in Cincinnati):

> Here I have been following now with earnest attention, for a few weeks, the practice of a very large London hospital, and I find that the majority of patients do get well; so I have come to this conclusion—that I must begin with a practice which is an old established custom, which has really more expressed science than any other system (the three heresies) but nevertheless, as it dissatisfies me heartily, I shall commence as soon as possible building a hospital in which I can experiment; and the very instant I feel *sure* of any improvement I shall adopt it in my practice, in spite of a whole legion of opponents. . . . I advise you E. to familiarize yourself with the healthy sound of the chest. I wish I could lend you my little black stethoscope that I brought from the Maternité.

When Elizabeth returned to New York she was too poor to realize the dream of building an experimental hospital. "If I were rich," she had told her sister, "I would not begin private practice, but would only experiment. As however I am poor, I have no choice." Choice absent, she went about doing God's work on earth. She set up a general practice and spent cold winters and steaming summers trudging the pavements with her black bag. Work on, Elizabeth! Proudly sporting the stethoscope she had brought from the Maternité, she attended to mainly poor, mainly female, chiefly immigrant patients. Her early years as this nation's first woman doctor of medicine were not encouraging. She confessed her deep unhappiness: "I had no medical companionship, the profession stood aloof, and society was distrustful of the innovation. Insolent letters occasionally came by post, and my pecuniary position was a source of constant anxiety." It was impossible to rent an office, the term "female physician" having been preempted by ill-trained abortionists, and she went into debt by buying a house on East Fifteenth Street. She worked in the attic and the basement, renting out the remainder of the house. Her isolation prompted her to adopt a 7-year-old orphan, Katharine Barry, and this young child became a lifetime companion, friend, and housekeeper.

Slowly, Elizabeth Blackwell began to attract support from the New York Quaker community, and in 1854 she opened on the Lower East Side a one-room dispensary in which she treated more than two hundred women in the first year. Its first annual report, written to rouse further support, spelled out its aims:

> The design of this institution is to give to poor women the opportunity of consulting physicians of their own sex. The existing charities of our city regard the employment of women as physicians as an experiment, the success of which has not been sufficiently proved to admit of cordial co-operation. It was therefore necessary to provide for a separate institution which should furnish to poor women the medical aid which they could not obtain elsewhere.

By 1856, the sisters were reunited. Emily had also had extensive medical training after having earned an MD degree from Western Reserve. She had walked the wards of Bellevue in New York and spent two *Wanderjahre* working with Sir James Young Simpson in Edinburgh, Franz von Winckel in Dresden, and William Jenner at the Children's Hospital in London. The Blackwell pluck and persistence paid off. Elizabeth's goal of founding an experimental hospital was achieved, but the medical experiments performed in it were in the field of women's rights and social justice rather than physiology or therapeutics. With the help of progressive philanthropists and their good friend Horace Greeley, the Blackwells established in 1857 the New York Infirmary for Women and Children at 64 Bleecker Street. They successfully overcame every objection of the time: that female doctors would require police protection on their rounds; that only male resident physicians could control the patients; that "classes and persons" might be admitted whom "it would be an insult to treat" (i.e., beggars and prostitutes); that signatures on death certificates might be invalid (the legal rights of women in the presuffrage era were fragile); that the male trustees might be held responsible for any "accidents"; and that in any case no one would supply women with enough money to support such an unpopular effort.

With Emily in charge of this going concern, Elizabeth traveled back to England and became the first woman to be registered as a physician in that country. She studied programs of maternal hygiene, looked over public health programs for women and children, and toyed with the notion of spending the rest of her life founding a country hospital with Florence Nightingale, to whom she had formed an intense personal attachment. The delicate overtones of her memoir anticipate lines that Katharine Lee Bates dedicated to Nightingale a generation later:

> Fragrant thy name as the City of Flowers;
> Sweet thy name as a song in the night;
> Over all wonders of womanhood towers
> Thy glory, white as the Cross is white.

When Elizabeth returned to New York in 1860, the sisters enlarged the infirmary, added new staff, and put in place the preventive measures of the sanitarian revolution. Their most critical innovations were in community health; they were the first to send "sanitary visitors" to the poorer neighborhoods of the city; their Tenement House Service was the earliest instance of medical social service in the United States.

The Civil War engaged the abolitionist spirit of the Blackwell family. On the day after Fort Sumter was fired on, the Blackwells helped to found the National Sanitary Aid Society; this became the nucleus of the Sanitary Commission. The Blackwells also acted as a conduit for the nursing corps which Dorothea Dix was assembling in Washington. Elizabeth wrote:

> All that could be done in the extreme urgency of the need was to sift out the most promising women from the multitudes that applied to be sent on as nurses, put them for a month for training at the great Bellevue Hospital of New York, which consented to receive relays of volunteers, provide them with a small outfit and send them on for distribution to Miss Dix.

When the war was over and finances permitted, another dream of Elizabeth Blackwell was realized. In the course of those long nights at the Maternité she had written, "What chance have women shut out from these instructions? Work on, Elizabeth!" It was another nineteen years before women could receive such instructions in New York: in 1868 the Blackwells founded a modern medical college for women, which by 1908, when it was absorbed by the newly coeducational Cornell University Medical College, had graduated 394 women doctors. Work on, Elizabeth, indeed! The faculty included such eminent female physicians as Mary Putnam Jacobi, considered the founder of pediatrics, and Elizabeth Cushier, professor of gynecologic surgery. The laboratories for instruction in both basic and applied sciences were among the most up to date in the country, and the three-year curriculum exceeded in rigor much of what passed for medical education elsewhere. Elizabeth Blackwell became the first professor of hygiene, and it was due to her efforts and those of Emily, who succeeded her, that many of the leaders of the sanitarian revolution got their start at the Women's Medical College of the New York Infirmary for Women and Children.

Perhaps the most remarkable aspect of the Blackwell family record is not their sanitarian or abolitionist zeal, but the large swath the family cut in feminist history. If, in the phrase of William James, the James family seemed to have "a serpent in its blood," the Blackwells had a colt. But their sanguine temperament ran in a direction opposite to that of the saturnine Jameses. Nowadays when one speaks of "the Blackwells" one includes not only Elizabeth and Emily but also a small tribe of reformers that spanned three generations. Of the twelve children of Samuel and Hannah Lane Blackwell, three died in infancy; Anna became a newspaper correspondent for Horace Greeley; Ellen developed into an author and artist; Emily and Elizabeth were doctors; and two of the brothers, Samuel and Henry, had public careers.

Samuel Blackwell married Antoinette Louisa Brown, who in 1853 was ordained as the first woman minister of a recognized denomination in the United States (Congregational). One of the few pulpits that offered her a guest appearance was that of the Universalist congregation in Worcester, Massachusetts, led by Thomas Wentworth Higginson—the Higginson who nurtured Emily Dickinson's verse and who became colonel of a black regiment of freed slaves in South Carolina. As Louisa Brown, the future Blackwell had worked in New York's prisons and hospitals and written accounts of their need for institutional reform that were published in Greeley's *Tribune* and later collected as *Shadows of Our Social System* (1856). Samuel Blackwell had sought her out at her upstate church precisely because of her writings. An abolitionist to the core, she became after the Civil War an ardent supporter of the suffrage movement, and both Blackwells were active in Julia Ward Howe's Association for the Advancement of Women.

Henry Browne Blackwell, himself an ardent abolitionist and Free-Soiler, married Lucy Stone, an Oberlin classmate of Louisa Brown Blackwell. A pioneer feminist, she insisted on keeping her own last name after marriage; women who adhered to this custom were for a while called "Lucy Stoners." At their wedding ceremony, on May 1, 1855, the Blackwells agreed on having read publicly a protest against the marriage laws then on the books. The protest was given wide circulation in an account by the minister who presided at the wedding: Thomas Wentworth Higginson. Henry, seven years her junior, had courted Lucy for two stormy years before finally winning her by saving a fugitive black woman from her owners. After the Civil War, Lucy Stone, Henry Blackwell, and their daughter Alice edited the influential feminist periodical the *Women's Journal.* The *Journal* became the official voice of the American Woman Suffrage Association, which Lucy Stone and Julia Ward Howe formed in 1869. The final public speech that Lucy Stone delivered was at the Columbian Exposition in 1893, naturally enough at the Woman's Building so beloved by Katharine Lee Bates. It was Stone's last chance at alabaster; she dodged the tomb and continued the Blackwell record of firsts by becoming the first person to be cremated in New England.

The personal affairs of Elizabeth and Emily Blackwell remained as monogamous as those of their brothers. Aside from her early involvement with Florence Nightingale, Elizabeth spent all of her life—the last thirty years in seaside retirement—with her adopted daughter and friend, Kitty Barry. Emily and her lifelong companion, Dr. Elizabeth Cushier, spent twenty-eight happy years together in a Gramercy Park brownstone in New York and on the coast of Maine.

It is difficult to find a group of men and women more enmeshed than the Blackwells in the great movements of the nineteenth century, to find a family more involved with intellectual conquest and social good. But many would argue that the Holmes family made as great a contribution to reform as the Blackwells. The two Oliver Wendell Holmeses, the Autocrat of the Breakfast-Table and his son, an Associate Justice of the Supreme Court, took the Brahmin road in support of many of the Blackwell causes, and the written record they left behind is more glittering by far than that of the Blackwells. Holmes Jr. thrice shed blood for the Union, and the doctor caused many to shed tears for the cause. In one of his "Medical Essays" of 1864 the elder Holmes sums up the political philosophy that he and the Blackwells espoused, and which his son was to make part of the common law:

> This Republic is the chosen home of *minorities,* of the less power in the presence of the greater. It is a common error to speak of our distinction as consisting in the rule of the majority. Majorities, the greater material powers, have always ruled before. The history of most countries has been that of majorities,—mounted majorities, clad in iron, armed with death, treading down the tenfold more numerous minorities. In the old civilizations they root themselves like oaks in the soil; men must live in their shadow or cut them down. With us the majority is only the flower of the passing noon, and the minority is the bud which may open in the next morning's sun. We must be tolerant, for the thought which stammers on a single tongue to-day may organize itself in the growing consciousness of time, and come back to us like the voice of the multitudinous waves of the ocean on the morrow.

It is now more than a century and a half later, and while the condition of women is not what it was then, we can see that the thoughts that stammered on the Blackwell tongue were slow in organizing themselves. It was not until 1915 that the medical school of New York University consented to give its first faculty position to a woman, Dr. S. Josephine Baker, one of this century's pioneers of social medicine. Dr. Baker had graduated from the Women's Medical College of the New York Infirmary for Women and Children in 1895, and after a distinguished period of practical work, she became a lecturer in child hygiene at Bellevue Hospital. In her autobiography, she wrote:

> They never allowed me to forget that I was the first woman ever to impose herself on the college. I stood down in a well with tiers of seats rising all around me, surgical-theater fashion, and the seats were filled with unruly, impatient, hardboiled young men. I opened my mouth to begin the lecture. Instantly, before a syllable could be heard, they began to clap thunderously, deafeningly, grinning and pounding their palms together. Then the only possible way of saving my face occurred to me. I threw back my head and roared with laughter, laughing at them and with them at the same time—and they stopped, as if somebody had turned a switch. I began to lecture like mad before they changed their minds, and they heard me in dead silence to the end. But the moment I stopped at the end of the hour, that horrible clapping began again. Frightened and tired as I was from talking a solid hour against a gloweringly hostile audience, I fled at top speed. Every lecture I gave at Bellevue, from 1915 through to 1930, was clapped in and clapped out that way; not the spontaneous burst of real applause that can sound so heart-warming, but instead, the flat, contemptuous whacking rhythms with which the crowd at a baseball game walk an unpopular player in from the outfield.

We've come a long way since 1930, and these days the impediments to full equality in the profession are more likely to arise de facto than de jure, but the verdicts of prejudice can be as stern as those of law. The medical and social causes to which the Blackwells, the Holmeses, the Stones, and the Bakers devoted themselves have by and large prevailed: abolition of slavery achieved, the Union preserved, sanitation promoted, infections curbed, child and maternal health protected by the state, women's rights in the profession moving slowly ahead. We are farther along than

Elizabeth Blackwell hoped for on her return from London. Women are not what they were over a century ago. Yet we are still some distance from realizing her fondest hope, that of a social movement which would unite the sexes under the banner of moral reform.

On December 1, 1850, Elizabeth Blackwell wrote that she regretted she had been unable to attend the Convention for Women's Rights held in Worcester, Massachusetts, the previous October:

> But I feel a little perplexed by the main object of the Convention—Women's Rights. The great object of education has nothing to do with woman's rights or man's rights, but with the development of the human soul and body. . . . My great dream is of a grand moral reform society, a wide movement of women in this matter; the remedy to be sought in every sphere of life. . . . Education to change both the male and female perverted character; industrial occupation, including formation of a priesthood of women; colonial operations, clubs, homes, social unions, a true Press; and the whole combined that it could be brought to bear on any outrage or prominent evil.

George Eliot could have called that grand moral reform movement "meliorism"—and she did.

17.
Call Me Madame

Le concret c'est de l'abstrait rendu familier par l'usage.
[The concrete is the abstract made familiar by usage.]

—Paul Langevin (1923)

On April 19, 1906, the 47-year-old Nobel laureate, Pierre Curie, was run over by an oversize, horse-drawn wagon filled with bales of army uniforms. He was attempting to negotiate the tricky Parisian intersection where traffic from the rue Dauphine, the quai de Conti, the quai des Grand-Augustins, and the Pont Neuf have created Gallic havoc for over a century. Curie had just left a meeting of reform-minded university professors at which he had argued for legislation to improve the lot of junior faculty and to prevent laboratory accidents. He had planned to stop at his publisher's office, but the office was shut because of a strike by equally reform-minded trade unionists. Absentminded and somewhat radium-sick, he turned away in the spring rain, and was on his way to the library of the Institut, when the six-ton wagon rumbled down the bridge from the Île de la Cité and crushed his skull.

His death brought to an end two remarkably creative careers in physical science, his own and that of his wife, Maria Salomea Sklodowskaa, known to the world as Mme Curie. On the rue Dauphine, she later recollected, "I lost my beloved Pierre and with him all hope and all support for the rest of my life." She was right: for although Mme Curie was to survive her husband until 1934, her contributions to science after his death were less than innovative; she turned her tough mind to the application of their discoveries, to teaching young scientists, and to construction of the Radium Institute, which she turned into a world center of physical science. She also became a secular saint of feminist culture on both sides of the Atlantic, the subject of her daughter's best-seller *Madame Curie* (1937), and heroine of sentimental publicists who hailed her for real or fancied radium cures. "More nonsense has been written about radium than the philosopher's stone," complained George Bernard Shaw in 1931, and he was right.

But there was no nonsense about the science. What a run the two Curies had together! In the course of six short years they had laid the foundations for the next century of physics and set the clock of our atomic age. That work earned Pierre and Marie Curie an acclaimed Nobel Prize in Physics (with Henri Becquerel in 1903) and Marie a more controversial Nobel Prize in Chemistry (1911). Those six years are the centerpiece not only of Eve Curie's biography of her mother, but also of all subsequent such works, including those of Françoise Giroud (1986) and Susan Quinn (1995). Perhaps in keeping with the temperament of their subject, each is written with more diligence than grace.

The Curies were married in 1895, after Pierre had already become famous for his work with his brother, Jacques Curie, on piezoelectricity (the phenomenon that some crystals, e.g. ceramic or bone, generate an electric current when compressed). He soon earned his doctorate for studies with Paul Langevin on paramagnetic resonance, establishing that the moment of an atom or electron varies inversely with temperature. It was the year that Wilhelm Roentgen took the first picture of the bones of his wife's hand by means of his novel rays.

By 1897, Henri Becquerel had found that uranium also produced rays—"emanations"—that left Roentgen-like shadows on photographic plates kept in the dark. Almost simultaneously, William Thomson, Lord Kelvin, discovered that the "ionizing" emanations from uranium imparted an electric charge to the air. In December of that year, Pierre and Marie set out to quantify the Becquerel emanations—the ionizing radiation—of a great variety of natural substances. For this purpose they employed the piezoelectric quartz balance, an instrument that Pierre had designed, and by February had found that the residue of pitchblende from which uranium had been extracted gave far greater signals than uranium itself. They correctly deduced that there was an ionizing substance far more active than uranium lurking in the sticky brew. It was the same year that Zola wrote "J'accuse" and France split forever into the Dreyfusards and their opponents.

By the end of 1898, the Curies had postulated that the new element, which they named "radium," decayed into another element, which they called "polonium." They gave the name "radioactivity" to the emanations from these elements. In 1902, by means of heroic preparative procedures, Marie Curie at last isolated radium in pure form. Later that year, Pierre calculated that one gram of radium emitted 3.7×10^{10} disintegrations per second: we call this amount of radioactivity one curie. And shortly thereafter he made the heuristic discovery that one gram of radium could heat one gram of water from 0° to 100°C: we call this sort of transformation "atomic energy," and today it powers more than half of France. By 1903 Pierre and Marie Curie had won a Nobel prize; they had also come down with the first signs of radium sickness.

Six unmatched years of discovery in the setting of the Third Republic, with axes drawn between right and left, church and state, theory and application, risk and benefit of a new science in a new century. It's a grand story, and while the Curies were on the spoor of the new, with the Dreyfus case breaking about them, it's an exemplary tale of science in service to reason. But after Pierre's death on the rue Dauphine, the story of Marie Curie becomes less a life in science and more a story of The Career, The Scandal, and The Legend. Her biographers (hagiographers?), led by Eve Curie, take us through laundered accounts of the Curie's adulterous affair with Paul Langevin, the outrageous attacks on her by the anti-Dreyfusard press, and the turn of her attention from science to the broader social scene. These proved to be as successful as her work in the lab. It was in recognition of the many mobile X-ray units she organized during World War I that a grateful France forgave her for the Langevin affair by permitting her to establish the Radium Institute.

It is difficult to guess what inner doubts or conflicts might have troubled the pale, intense widow in a plain black dress who lived on the fashionable quai de Béthune, the *entrepeneuse* who raised millions in France and the United States for her Institute by encouraging claims of cures for cancer. Nor does any material yet published yield insight into what must have been the remarkable relationship between Mme Curie and her daughter, a physicist at her mother's Institute, who married a brilliant young coworker to play out the story of *Marie et Pierre Redux.* Irène and Frédéric Joliot-Curie not only shared a Nobel Prize (Chemistry, 1935) "in recognition of their synthesis of new radioactive elements"—the third Nobel in one family—but also an abiding attachment to Soviet communism. The story of the Curies reached from the quai de Béthune to the podium of the Comintern.

The political and dynamic undertones of this part of the story are not in the public record. But that would have been just fine with Mme C: except for some painful letters addressed to her husband after his death, her private voice was as impersonal as her public speech. The Curies met in 1894 and were made for each other. Both Maria Sklodowka, daughter of a Polish gymnasium teacher, and Pierre, son of a Communard homeopath, were raised in the frugal folkways of the hardworking petite bourgeoisie. One catches the flavor of Curie's biographers—and a quaint view of genetics—from Susan Quinn's description of how Marie's mother learned to cobble her children's shoes: "Such willingness to do manual work was inherited by her youngest daughter, and was essential to Marie's success many years later in isolating radium." The intensity with which the Curies stare at us from their public portraits suggests that what Einstein said of Marie might have applied to Pierre as well: "Madame Curie is very intelligent but has the soul of a herring, which means she is poor when it comes to the art of joy and pain."

The two cold fish had sought each other out from among a flashier school of broadly cultivated, anticlerical scientists collected by the Sorbonne at the *fin de siècle.* Those glittering mathematicians, physicists, and chemists formed the phalanx of the positivist movement and became a vanguard of the Dreyfusards. As might be expected, the Sorbonne positivists became the targets of the protofascist right and the fans of *La France profonde*, the deep, "true" France. As part of a successful attack on Marie Curie's nomination to the Académie, Leon Daudet chimed in against the professors in his rightist journal *l'Action française:*

> They are all like that fanatic Poincaré, a man of genius, they say, in mathematics but stupid and hateful. When it comes to the rest; the Jew of color photography Lippmann, that fanatic Dreyfusard Appell, dean of the faculty of Sciences . . . [they] no longer hide behind the life of the Saints, but behind algebra, physics and chemistry treatises. . . . They intend in fact to chase from the house all who don't think like them, don't feel like them, who have the audacity not to deny God, not to insult Rome, go to mass, to raise their children as Christians.

Nowadays the Curies are ignored in cultural histories of the Third Republic (Jerrold Siegel's *Bohemian Paris* and Eugen Weber's *Fin-de-Siècle Paris* come to mind). But from the laws and units with which their names will forever be associated, their discoveries remain part of the fabric of twentieth-century science. For me, the lives and work of these Dreyfusards of science constitute a monument to reason that their contemporaries in the arts have not quite matched. Official France agrees: their names are woven into the fabric of Paris. The square before the École Polytechnique (in the fourth arrondissment) is named after the dashing but very married Paul Langevin, whom Einstein said would have discovered relativity had he himself not done so and whose affair with widowed Marie almost cost her the second Nobel prize. His was the noblest career in the resistance—and he suffered for it. The square before the Sorbonne (also in the fourth) is named after mathematician Paul Painlevé, who was Langevin's second at the duel he fought to preserve Mme Curie's honor. Painlevé was another early supporter of Einstein; he was political enough to become a minister of war. Langevin quipped that Painlevé had studied Einstein thoroughly, though unfortunately not until after he had written about him, a sequence acquired perhaps in politics. Marie's other partisans in the Curie–Langevin scandal dot the landscape as well: Gabriel Lippmann, the Nobel physicist who invented color reproduction and who presented the Curies' discovery of radium to the Academy of Sciences is remembered behind the Place de la Nation (in the twentieth arrondissement); Paul Appell, dean of the Faculty of Sciences, has an avenue of his own near the Cité Universitaire (in the fourteentth)—it leads to the avenue Rockefeller; the rue Henri

Poincaré loops off boulevard Gambetta (in the twentieth); and Emil Borel is off the boulevard Périphérique in the seventeenth.

The story of Pierre and Marie Curie is a tribute to a dazzling set of discoveries jointly made by a man and a woman of genius. *"L'Art c'est moi; la science c'est nous,"* wrote Claude Bernard, and that *nous* remains independent of gender, psychic baggage, or family romance. Among the most memorable photographs in Quinn's book is a late one of an intense Marie Curie on the balcony of her Institute behind the École Normale. Her lined face looks forward to the future, her hands are scrumbled by the scars of radium; it's an image that sums up the hope and the harm of her discovery. She would have been pleased that she is shown overlooking the street that we now call the rue Pierre et Marie Curie.

Beside the Golden Door

18.

Welcome to America: Einstein's Letter to the Dean

More than 4 million Syrian refugees have been forced to flee the country. . . . The United States has a long history of helping the world's most vulnerable people, but we have also faltered when faced with difficult decisions to allow refugees into the country.

—Message to the President from seventy-two Members of Congress (September 2015)

We can delay and effectively stop for a temporary period of indefinite length the number of immigrants into the United States. We could do this by simply advising our consuls to put every obstacle in the way [to] postpone and postpone and postpone the granting of the visas.

—Breckinridge Long, U.S. State Department (June 26, 1940)

Please go at once to the Dean of that University, Mr. Currier McEwen, and give him the enclosed letter.

—Albert Einstein (August 10, 1939)

IN THE FALL OF 2015, THE NEWS WAS ALL ABOUT BORDERS. On the one hand, Médecins Sans Frontières (Doctors Without Borders) won a well-deserved Lasker Award for their response to the Ebola virus; on the other hand, Hungary erected a border-long barbed-wire fence to prevent tattered Syrian refugees from tracking north. Transgressors were treated to pepper spray. Not to be outdone, American nativists demanded a thirty-foot-high barrier across our southern border to deter Mexican immigration and a somewhat lower barrier on our northern border to exclude the riffraff from Canada.

After an early welcome by Germany, borders closed across Europe to leave the bulk of Syrian refugees camped in unplowed fields in Greece or Serbia. Hunger, thirst, and dysentery were rife among the huddled masses yearning to breathe free farther north. Those who managed to breach the Hungarian border were packed into buses destined for reception camps that promised freedom and a shower. Viktor Orban, prime minister of Hungary, claimed that he was protecting European civilization. Slovakia would admit only Christian refugees, not the Muslims of Syria or Afghanistan. At a Czech railway station, officials detained refugees to scrawl identification numbers on their forearms with felt-tipped pens.

Promises of freedom after a shower, racial quotas, and indelible numbers on the forearm seemed like flashbacks to the Hitler years. Newsreels of the worst episode in Central European history were resurrected on iPads and laptops in 2015. It is no wonder that memories of the Holocaust prompted members of Congress to remind an American president that our country had "faltered when faced with difficult decisions to allow refugees into the country."

"FALTER" SEEMS A TAD MILD IN RETROSPECT. Immigration policies set by a nativist U.S. Department of State in the late 1930s assured that 90 percent of the quota slots allotted for immigrants from countries ruled by Hitler and Mussolini would remain unfilled. It has been estimated that had those visaless refugees received consular approval, an additional 190,000 could have escaped the teeming shores of Europe—or the showers at Auschwitz. Alas, the director of the Immigrant Visa Section at the Department of State,

Breckinridge Long (1881–1985), wasn't lifting any lamps beside the golden door. Long and company worked out a simple solution to limit immigration of refugees, mainly Jewish: paperwork, paperwork, and more paperwork. A paperwork fence would protect our borders!

Long had ordered consular officials in Europe "to postpone and postpone and postpone the granting of visas." One assumes that Long's previous experience as ambassador to Mussolini's Italy had guided his policies. In 1933, in a letter to Joseph E. Davies, ambassador to the Soviet Union, he had praised the Fascist regime in terms that Ezra Pound could not have bettered, and to Almy Edmunds, wife of Judge Henry L. Edmunds, he wrote, "Italy today is the most interesting experiment in government to come above the horizon since the formation of the Constitution 150 years ago today."

Breckinridge Long and his paperwork fence set the stage for an urgent letter by Albert Einstein to Dean Currier McEwen dated August 10, 1939. Einstein was worried that Professor Rudolf Ehrmann, who had been Einstein's personal doctor in Berlin, would be swept up in a Nazi roundup. However, between Ehrmann and a visa to America stood the paper fence of Breckinridge Long. Although the doctor had already gained a year-long lecture appointment at New York University, the U.S. consul dragged his legs and demanded written assurance that Ehrmann receive a full-time two-year appointment at the school. The date of the letter, August 10, 1939, is telling—eight days after the famous Einstein–Szilard letter to President Roosevelt explaining that it had "become possible to set up a nuclear chain reaction in a large mass of uranium, by which . . . extremely powerful bombs of a new type may thus be constructed." Four weeks later, Hitler invaded Poland, and World War II began.

EHRMANN WAS NOT YOUR AVERAGE FAMILY DOCTOR; he had left his mark on medical science. Physiologists remember his bioassay on adrenaline (based on dilation of the amphibian pupil). Gastroenterologists know his work on pancreatic function and gastric acid secretion, and rheumatologists of my generation recognize his name in the Ehrmann–Sneddon form of a skin condition known as livedo reticularis. Ehrmann had risen to the rank of professor of internal medicine at the University Hospital of Berlin, and his hometown of Altenstädt named him one of the 50 most illustrious Germans.

However, in August 1939, there he was: a 60-year-old man huddled on the teeming shores of *Heimat*—"homeland"—with the Gestapo in the wings. Einstein received an urgent appeal from Ehrmann's son Rolf, who himself had escaped from Germany the year before. In that note, Ehrmann asked Einstein to write directly to Washington, an effort that the physicist knew to be completely fruitless "as I know from previous experience." Einstein suggested a more receptive ear than that of Breckinridge Long:

> We must try to get Prof. E. a fellowship for 2 years at NYU. Please go at once to the Dean of that University, Mr. Currier McEwen, and give him the enclosed letter. If he is away, ask for his address or find out if someone else can make such an important decision in his stead. Tell him that a 2-year contract is required by the consul. Tell him, too, that if the University is so inclined, we have a good prospect of having his salary for the second year guaranteed by an immigration organization.

McEwen responded immediately. It was already August, but the appropriate letters went back and forth rapidly, and by the time war had begun, Ehrmann was on his way to the United States. By October, he had taken his place as a clinical professor of medicine (his specialty, gastroenterology) at the NYU School of Medicine and as an attending physician at Bellevue Hospital. He went on to establish a clinical practice based at Beth Israel Hospital and published a half a dozen articles in English before his retirement. And in 1955, when Einstein was terminally ill, Ehrmann rushed down to Princeton, New Jersey, to be by his side.

McEwen's almost reflex action was one of a series of humanitarian gestures that served him and his university well. Rockefeller-trained, Currier McEwen was a distinguished rheumatologist before the term was even invented. He had been a founder and president of the American

Rheumatism Association (now the American College of Rheumatology) and made important contributions to the treatment of rheumatic fever. As dean of the NYU School of Medicine from 1937 to 1955, he presided over at least twenty lifesaving appointments to German and Austrian medical scientists. Among these were the radiologist Rudolph Bucky, the biochemist Ephraim Racker, and Nobel laureate Otto Loewi.

Loewi's story is almost an echo of Ehrmann's. Otto Loewi, professor of pharmacology at the University of Graz in Austria, had won a Nobel prize in 1936 for his discovery of acetylcholine, known in German as *Vagusstoff.* But no matter how high the honor, when the Nazis seized Austria in 1938, he was awakened from his sleep by a dozen young storm troopers armed with guns who hustled him off to jail. Loewi owed his rapid release from prison and temporary refugee status in London to his co-holder of the 1936 Nobel Prize in Physiology or Medicine, Sir Henry Hallett Dale. However, with no job and no money, Loewi had been forced to instruct the Nobel organization's bank in Stockholm to transfer the Nobel prize money to a prescribed Nazi-controlled bank. Once safe in England, Loewi asked his colleague, Walter B. Cannon (the *Wisdom of the Body* Cannon), to lobby for a position at Harvard University. However, by 1940, an influx of great and near-great European scientists to Harvard had made President James Conant reluctant to offer a faculty position to other immigrants, even to a Nobel laureate.

Cannon found another solution. He persuaded Homer Smith, professor of physiology at NYU School of Medicine, to offer Loewi a full-time professorship in pharmacology; Smith had been a research fellow in Cannon's laboratory from 1925 to 1926. Dean McEwen chimed in, funding was obtained, and Loewi arrived for work in 1940. When Cannon visited him in New York, Loewi expressed his heartfelt "admiration for American freedom and the generous attitude of the American people for refugee scholars." In 1952, when Otto Loewi gave a pharmacology lecture on his frog-heart experiments to my medical school class, Rudolf Ehrmann, his faculty colleague, was sitting in the front row.

Loewi was the first of three Nobel laureates to have worked at NYU School of Medicine; each was foreign born. The others were Severo Ochoa (recipient of a Nobel prize in 1959 for work on RNA polymerase), who came as a refugee from Loyalist Spain in 1942, and Baruj Benacerraf (recipient of a Nobel prize in 1980 for work on immunogenetics), who arrived in 1956 after years in Paris and the American boondocks. Today, their students fill the halls of academe and the laboratories at Woods Hole where Loewi, Ochoa, and Benacerraf established a yearly beachhead at the Marine Biologic Laboratory. It is where Loewi told students that "a drug is a substance which, if injected into a rabbit, produces a paper."

A RECENT STUDY BY THE VILCEK FOUNDATION (Fig. 1) found that of the 277 Nobel prizes awarded to scientists working in the United States from 1901 to 2014, 35 percent went to the foreign-born.

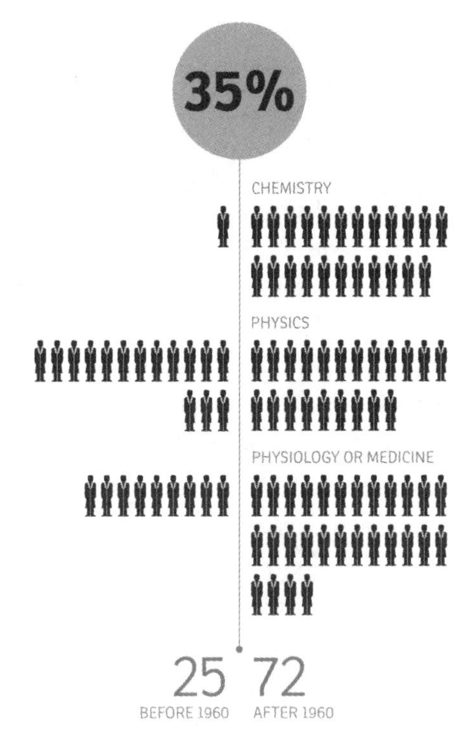

Fig. 1. Foreign-born American Nobel laureates before and after the Immigration and Nationality Act of 1965 relaxed quota restrictions. (The Vilcek Foundation)

However, the foundation argues that the statistics are "end loaded." Almost 75 percent of those Nobel prizes were awarded only after the Immigration and Nationality Act (the Hart–Celler Act) was passed in 1965, eliminating discriminatory national-origin quotas and increasing employment-based green cards. This permitted a dramatic increase of the Asian population in America from less than 1 million in 1960 to more than 17 million in 2010, becoming approximately 5 percent of the U.S. population. Science got more than its share: of those 72 post–1965 Nobel laureates, 13 were Asian-Americans; that is 17 percent of U.S. laureates, and three times the percentage of Asians in America. Biologists will recognize Har Gobind Khorana (Nobel prize in 1968 for work on the genetic code) and Roger Tsien (Nobel prize in 2008 for work on green fluorescent protein).

Asian-Americans and others who come to the United States for pre- and postdoctoral studies with hopes for a green card are a human resource in science that has been indispensable to the success of the enterprise. A random issue of the *FASEB Journal* (December 2015) can serve as an example of international contributions to the field of biology. It contains 27 articles, of which 11 are from the United States—a proportion that has held steady for a decade. Those articles list a total of 69 authors, 35 of which have Asian surnames. Any quick survey of other biomedical journals will confirm this ratio. Without this influx of immigrant post-docs, biomedical science would be the poorer.

Signing the Hart–Celler Act on October 3, 1965, at Ellis Island, President Lyndon Johnson promised that, from that day onward, we would welcome immigrants "because of what they are and not because of the land from which they sprung." So now, let's revive Ellis Island as an entry point and welcome to America some of those ragged refugees from the shores of the Mediterranean. It would honor the promise of Emma Lazarus inscribed on the base of the Statue of Liberty:

> Give me your tired, your poor,
> Your huddled masses yearning to breathe free,
> The wretched refuse of your teeming shore.
> Send these, the homeless, tempest-tost to me,
> I lift my lamp beside the golden door!

19.

Modernism and the Hippocampus: Kandel's Vienna

> "'The Scream' is more than a painting, it's a symbol of psychology as it anticipates the 20th-century traumas of mankind."
>
> —Ivor Braka in the *New York Times* (May 12, 2012)

Eric Kandel, perhaps the pre-eminent neurobiologist of our day, proves in *The Age of Insight* that he is also a cultural historian with a fine eye, a good ear, and—in his words—"a heart that dances in three-quarter time." No wonder that, although his family escaped Hitler to come to New York, he has never forgotten his "city of dreams."

Through the lens of Vienna 1900, a city on the brink of the modern age, Kandel explores the science of the brain today, explaining that we're on the brink of understanding emotional responses to the visual arts. He introduces us to the new science of neuroesthetics, in which cognitive psychology, brain imaging, and psychopharmacology play complementary roles in working out how vision, emotion, and empathy shape our experience of art.

The salons and cafes of Vienna 1900 were home to transgressive artists such as Oskar Kokoschka, Egon Schiele, and Gustav Klimt; louche writers such as Hugo von Hofmannsthal and Arthur Schnitzler; masters of new design and architecture such as Josef Hoffman and Otto Wagner;

and probing medical intellects such as Carl von Rokitansky and the great anatomist Emil Zuckerkandl. Vienna also spawned an influential school of art history with Alois Riegl, Ernst Gombrich, and Ernst Kris—and then, of course, there were Richard von Krafft-Ebing and Sigmund Freud. Many of them were entertained at the fashionable salon presided over by Berta Zuckerkandl, wife of Emil.

From this matrix, Kandel has precipitated a work that bids fair to rank with Arthur Koestler's magisterial *The Act of Creation* as a classic analysis of how art and science inform each other. In his volume, superbly illustrated by masterpieces of the movement, Kandel suggests that Viennese modernism had three major themes.

The first theme is that the human mind is largely irrational by nature and that if unconscious conflicts were brought to the surface, they would produce "new ways of thoughts and feeling." These new ways, often explicitly erotic or aggressive, would, as it were, get to the bottom of things. Kandel illustrates this point in a striking comparison of two paintings. He contrasts the reclining nude Venus in a sixteenth-century painting by Giorgione with a sprawling nude woman drawn by Egon Schiele. In the Renaissance painting, the gently curving hand of the goddess slopes to cover her mons pubis; in Schiele's drawing, a knees-up bawd is masturbating.

The second theme, writes Kandel, is the introduction of self-examination as a means of transcending outward appearances. In the search for the dynamics of self, the seeker would discover rules that govern human individuality. In addition, the artist's engagement with the self would evoke an empathic response from the viewer, a "vicarious experiencing of the subject of a painting." Kandel presents several portraits by Schiele and Kokoschka in which the artist's own visage is projected onto the face of the subject. Each portrait by a Viennese self-examiner is in fact a self-portrait:

> One reason for our emotional and visual response to faces in art is the important role that face perception plays in social interactions, emotions, and memory. Indeed, face perception has evolved to occupy more space in the brain than any other figural representation.

The third theme is the integration of human knowledge. Vienna opened a dialogue among biological sciences and psychology, literature, music, and art, and thereby "initiated an integration of knowledge that we are still engaged in to this day. It also transformed science in Vienna, especially medicine." Berta Zuckerkandl's anatomist husband introduced Klimt to descriptive biology in general and to Darwinian theory in particular. Klimt learned these facts of life well. His iconic Byzantine portrait of Adele Bloch-Bauer (these days as popular as the "Mona Lisa") is filled with enough elliptical ova, square spermatids, and major endocrine organs to fill a gynecology text. No matter: it's glorious. If you would seek a monument to the unity of arts and science, look no further than the many college dorm rooms over which Adele presides.

These three themes fill roughly one-third of Kandel's engaging work. The remainder of *The Age of Insight* is devoted to bringing into confluence two other major intellectual efforts that have engaged us since Vienna 1900. Art historians and cognitive psychologists have tried to explain the subjective response to a work of art that Gombrich called the "beholder's part." On the other hand, neurobiologists and psychiatrists have aimed to understand the objective forms and function of memory and desire. This part of Kandel's book, written by a master of the craft, is both an introduction to neurobiology and a critical review of what neuroesthetics is all about.

We learn that ancient applications in two regions of the inferior temporal cortex send visual information to three target sites in the brain: the lateral prefrontal cortex for downloading and filing the image; the hippocampus, which holds it on the memory drive; and the amygdala, which sends the up- or down-beat for the soundtrack. We get genetics, esthetics, history, and the latest in kinetic imaging. Any book that can go from neatly outlining the work of Kuffler, Hubel, and Wiesel on "figural primitives" to brisk analyses of recent observations by Larry Squire, Nancy Kanwisher, and Jonathan Cohen is broad indeed. But then again, we're in the hands of the man who, with Jim Schwartz and Tom Jessel, literally wrote the book on neural science.

But Kandel's *The Age of Insight* (and his experience) also touch on somewhat darker aspects of Hapsburgian Vienna. He concludes that Freud, Schnitzler, and artists of Vienna 1900 had explored in their work

> the ubiquitous presence of sexuality and its onset in early childhood; the existence in women of an independent sexual drive and erotic life that is equal to that of men; the pervasive existence of aggression; the continuous struggle between the instinctual forces of sexuality and aggression and the resulting anxiety to which this conflict gives rise.

Some may quibble that what's missing in *The Age of Insight* is a discussion of why Ernst Gombrich, Stephen Kuffler, Ernst Kris, and Sigmund Freud—not to speak of Eric Kandel—could not live out their lives in Vienna, but that's another story. Perhaps all of that unedited self-examination, aggression, and anxiety did more than just anticipate "the 20th-century traumas of mankind," to quote the *Times* on Edvard Munch. The explanation must be stored somewhere between the hippocampus and the amygdala. That brings us to more neurobiology.

KANDEL HAS SUGGESTED THAT THE MIRROR NEURONS in the motor areas of the brain "make us perceive the actions of others as if they were our own," and proposed that that's how the "beholder's part" is wired. I'd disagree. A trip to a Claes Oldenburg exhibit at the Museum of Modern Art in New York suggested to me that an equally important group of cells, the "grid cells," might also come into play in our experience of art. The exhibit was filled not only with the expected Oldenburg images of everyday objects, like forks and sneakers transformed into caryatids or crucifixions, but also huge plaster constructions of hot dogs, chocolate cakes, and cream pies. The mirror-neuron theory needs some help here. These room-size objects made me scramble around the exhibition like a mouse in a maze that the artworks defined. My grid cells were firing to help me work my way about: checking the brain's GPS.

Kandel is probably correct in defining a major role of mirror neurons when it comes to working out Kokoschka et al., but not for a lot of art from mid-twentieth century on. Think of all those "installations," "assemblies," and "tableaus" that fill galleries and museums these days. Many such works fill spaces that demand circumnavigation, like Oldenburg's cream pie. Their placement in space defines our map of attention, and so we turn to the brain's GPS: the grid cells.

And as for Oldenburg, I'd say that beauty is in the pie of the beholder.

20.

A Taste of the Oyster: Jan Vilcek's *Love and Science*

All art is autobiographical. The pearl is the oyster's autobiography.

—Federico Fellini (1965)

Were Jan Vilcek's book *Love and Science: A Memoir* just an autobiography, it would present itself simply as another American immigrant success story: A bright middle-class boy survives the Nazis in Czechoslovakia after family conversion to Catholicism. He obtains his medical and research education under the constraints of a Communist regime and begins laboratory work on viruses and interferon. He marries a beautiful art curator and both escape to New York, where he continues to study interferon until he comes up with a chimeric antibody to tumor necrosis factor, a drug called infliximab. The antibody doesn't cure tumors but works like gangbusters in autoimmunity. Infliximab is marketed commercially as Remicade and becomes a $10 billion a year drug for treatment of rheumatoid arthritis and Crohn's disease. Profits are used to fund major philanthropic efforts and patronage of the arts. Our hero receives the National Medal of Technology and Innovation from President Obama in 2013.

But Jan Vilcek's engaging book is not just an autobiography. It's the very model of the memoir as a literary genre.

If, following Fellini, a pearl is the oyster's autobiography, a memoir describes how the oyster tasted. In a memoir, personal experience lends flavor to historical fact, and Vilcek's *Love and Science* gives us both. I'd place his gentle, scholarly reflections on a small shelf with those of other fine scientists who have given us a taste of their oyster: Peter Medawar in *Memoirs of a Thinking Radish*, Francis Crick in *What Mad Pursuit*, and Erwin Chargaff in *Heraclitean Fire*.

The facts are tough reading; they include the stormy history of what happened to the Czechoslovakia of Vilcek's childhood. Forged from the embers of the Hapsburg Empire, the map of Czechoslovakia between 1919 and 1939 looked like a pancreas pushed into a fat, liver-shaped Germany. The Nazis lopped off the juicy Sudentenland in 1938 to provoke World War II, and the Slovaks split off a new state a year later. Nazi rule followed until the end of the war, and the Soviet coup of 1948 shifted the map again; Vilcek's book spells out the geography and chronology of those perilous years. The names changed, regimes were overturned, and the map of Central Europe turned cystic. Jewish families were threatened, fragmented, or—like Vilcek's grandmother—sent to the camps to perish.

In a section called "A Tumultuous Childhood," Vilcek tells us how at age 8 he was separated from his parents and placed in a Catholic orphanage, then hidden in the provinces with his mother: "We did have the feeling that they viewed us with some suspicion, because despite my mother's kerchief and my worn clothes, neither one of us looked as if we belonged there." Dodging Wehrmacht soldiers from the west and the Russians from the east, the family was reunited with their absent father at the end of the war: "My mother and I were standing by the highway contemplating what to do next, when, only a few minutes later, a civilian passenger car stopped, seemingly for us. The door opened and out stepped father."

Vilcek eventually entered medical school in Bratislava's Comenius University and was soon attracted to its Institute of Virology, where visiting dignitaries such as Alick Isaacs (interferon) and Macfarlane Burnet (clonal selection) lit the fires of ambition. Tough slogging at first; about a year and a half after joining the Institute he asked himself, "Why am I wasting my time here? I have a medical degree. I could be doing something more productive!" He applied for a clinical appointment, was put on hold, and plugged away at his painstaking work on tick-borne encephalitis virus. The work paid off. In July 1960, just a few weeks after his 27th birthday, three years after graduation from medical school and two years before his PhD, he published his first, single-authored paper in *Nature*, "An Interferon-like Substance Released from Tick-Borne Encephalitis Virus-Infected Chick Embryo Fibroblast Cells." More and more interferon papers followed, chiefly in the house organ of the Institute, *Acta Virologica.* Soon acknowledged by others in the field, he was permitted a short "study visit" to Louvain and London in 1962. At the National Institute for Medical Research at Mill Hill, he presented his work to Alick Isaacs and other mavens of British immunology, beginning his seminar by saying, "I feel like a country vicar who comes to Rome to lecture about the Bible to the pope and his cardinals."

But although the Soviet authorities denied him a chance to spend a post-doc year at Mill Hill, that brief visit was enough to open his eyes to the West. Defection loomed, and he'd acquired a new helpmeet to speed his way. He had met, matched, and married the statuesque Marica Gerháth, an assistant curator at the Slovak National Gallery. She was stunning in her New York wardrobe, faithfully shipped to her by her elder brother Ivan, who had defected to the United States a few years earlier. Marica became the "*Love*" of his memoir's title—and his life. The young couple soon planned defection. "I am too tall for this country," Marica would say, and Jan agreed—on professional grounds—that "if the opportunity to leave ever arose, we would get out."

In October 1964, the opportunity arose. The authorities permitted the Vilceks to accept an invitation from friends in Vienna to join them at the State Opera ("Can you bring your tuxedo?" their friends asked.) The young couple packed all their wordly goods (no tuxedo) into a small Skoda automobile and crossed to the West. Bratislava is only thirty-five miles from Vienna, but in those days the cities were separated by "watchtowers, minefields and electrified wire fences: the Iron Curtain." Their papers checked out, and the Vilceks crossed into the West, where the next evening was spent at a performance of *The Magic Flute*: "Listening to the lovely music," Vilcek remembers, "there were moments when I almost forgot the profound change we had made in our lives that day."

In time, after tedious consular visits and visa and passport hang-ups, Jan and Marica arrived in New York on February 4, 1965. Jan had been given a faculty appointment in microbiology at NYU School of Medicine, earned though the promise of his interferon research and the good offices of his brother-in-law Ivan, by then an anesthesiologist at University Hospital.

In New York, Jan was given an empty lab that previously had been used as animal quarters for the Medical School. He began to write grants, and wait. But soon he was funded, and the lab filled with machines and people. The Vilceks settled into junior faculty digs, and Marica started a new job in the Catalogue and Acquisitions Department of the Metropolitan Museum of Art. Both careers took flight. "And one day, in our third or fourth year in New York, Marica and I made a radical decision. From that day on we would no longer speak to each other in Slovak . . . only in English."

Love and Science continues the story of Vilcek's career as a productive investigator, a tale familiar to any scientist with an "h factor" over 50. It describes the flowering of his interferon work in the springtime of New York immunology. At NYU, Lewis Thomas, Sherwood Lawrence, John David, and Ed Franklin had drawn neat distinctions between humoral antibodies and factors released by activated lymphocytes, like macrophage inhibitory factor, or MIF. Dudley Dumonde named these factors "lymphokines"—the name soon to become "cytokines." At Albert Einstein College of Medicine, Barry Bloom and Boyce Bennett worked out the mechanics of lymphokine induction. At Rockefeller, Anthony Cerami and Bruce Beutler had isolated cachectin, soon found to be identical to tumor necrosis factor (TNF) and first isolated across York Avenue by Lloyd Old at Memorial Sloan Kettering. Vilcek wrote, "The fact that I was friends with Lloyd Old very likely influenced my decision to look for TNF in the materials we generated when producing IFN-gamma in my laboratory."

Vilcek moved briskly along, distinguishing leukocyte from fibroblast interferon, learning en route that human mononuclear cells are easier to handle than baby foreskin cultures. Then came "immune interferon"

(IFN-gamma). The molecule wasn't successfully cloned (cDNA) until 1982, but Vilcek's lab soon found that culture medium of stimulated mononuclear cells contained cytokines other than interferons. One of these was TNF: "Having worked on various aspects of interferon for over 20 years, I felt that perhaps it was time to look for inspiration elsewhere. Why not TNF?"

THE FIRST QUESTION WAS WHETHER TO EXPLORE the biology of TNF itself or the effects of blocking its actions. TNF was ineffective against human cancers, and antibodies to TNF failed trials in human sepsis, despite animal studies that supported these indications. But one of Vilcek's tailored antibodies to TNF (called cA2) worked well in autoimmune diseases such as rheumatoid arthritis. In these ailments, a brew of cytokines cause inflammation, traditionally defined as "redness and swelling with heat and pain," and cA2 blocked the critical cytokine: TNF.

Love and Science tells the story of Vilcek's trek from his "experiments with interferon to cA2, yet . . . the path was not always clear." As early as 1984, Vilcek had reached a licensing agreement between NYU and a biotech venture, Centocor: the Vilcek lab would provide the company with diagnostic monoclonal antibodies against interferons and other cytokines such as TNF. In turn, Centocor agreed to support the lab (eventually for fifteen years!) and to pay royalties to NYU for any sales. The emphasis soon switched from diagnosis to treatment. By 1988–1989, the mouse antibody to TNF (called A2) was formulated by Jimmy Le in the Vilcek lab. Working together with Centocor scientists, the Vilcek lab turned the mouse antibody into a chimeric mouse-human antibody (that now famous cA2).

Vilcek recalls milestones in the development of cA2 into the clinical drug Remicade (infliximab). Formulated in 1988, the drug first gained FDA approval for Crohn's disease in 1998. Ten years from bench to clinic merits the modest claim of "no small accomplishment." Remicade (produced by J&J/Janssen) is now also approved for rheumatoid arthritis, psoriatic arthritis, ulcerative colitis, and other disorders, and by 2014 three million patients worldwide had received the drug. It ranked among the top three drugs in the marketplace, with sales close to $10 billion a year. Some of this came back to First Avenue: Vilcek estimates that "Royalty payments collected by NYU . . . by now will have exceeded $1 billion." More passed through to the inventor, permitting him to "acquire material wealth that I had never aspired to, catapulting me into a new world of philanthropy and art collecting." He soon became at home in that new world.

FOLLOWING THE EXAMPLE OF ANOTHER immigrant philanthropist, Andrew Carnegie, the Vilceks report that they have "given away an amount that is much greater than all of our personal assets combined." The major recipients of that largesse were venerable institutions, such as Jan's NYU and Marica's Metropolitan Museum of Art. But by 2006 the Vilceks had launched a new effort, cash-rich annual prizes to recognize contributions made by American biomedical scientists and artists born abroad. Jan tells us, "Both of us are immigrants. I am a biomedical scientist and Marica is an art historian. Could we perhaps build a foundation program around combined experiences?"

They did, and over the last decade the Vilcek prizes have featured a dazzling cast of recipients: from Rudolph Jaenisch (stem cells) to Mike Nichols (film), from Titia de Lange (telomeres) to Michael Baryshnikov (dance), from Richard Flavell (immunology) to Charles Simic (poetry), and more. Awards of promise have also been given to young tyros in the field, and a new foundation building has opened gallery space, filmmaking facilities, and research space in which to document the contributions of immigrants to art and science in America.

A final note: there cannot be many scientific memoirs the author of which is familiar with the work of both polio pioneer Albert Sabin and modernist painter Marsden Hartley, of rheumatologist Ravinder "Tiny" Maini and the doyen of *Wiener Werkstätte*, Josef Hoffman. The last chapter of Vilcek's memoir recounts the success of the Vilcek collection of American Modernist art and Native American artifacts, a collection meant to endure; its exhibition has traveled from Manhattan to Oklahoma. I'd say the same for *Love and Science*; its author is meant to endure, and the book is ready to travel worldwide. It's a pleasure to read; indeed, we might conclude that for Jan Vilcek, the world is his oyster.

21.

Richard Dawkins Lights a *Brief Candle in the Dark*

Brief Candle in the Dark is the most engaging—perhaps the best—book of the baker's dozen or so that Dawkins has written. Subtitled *My Life in Science*, it's billed as the second volume of his autobiography, but it casts a net far wider than its predecessor, *An Appetite for Wonder*. Volume 1 gave us his bio basics: an upper-class English childhood in Kenya, a coed independent school (Oundle) in Peterborough, then Balliol College, Oxford, followed by an Oxford MA and PhD with Nikolaas Tinbergen (Nobel Prize in Physiology or Medicine, 1973). He went on to a stint at Berkeley in the days of pot and protest. With stickers of Gene McCarthy's presidential campaign on his Ford Falcon station wagon, he drove cross-country and returned on the SS *France* to Oxford for good. There he planted his foot on the shores of modern biology with *The Selfish Gene* (1976) and for over a generation, with wit and panache, has carried the news of Darwinian evolution to a general audience.

In *Brief Candle in the Dark,* Dawkins brings his own contributions up to date, hoping that "taken together these might add up to a kind of biologist's worldview, with an aspiration at least to coherence." His formulations add up to more than a worldview, more than a weltanschauung like Immanuel Kant's; he has presented testable concepts of daily use to experimental biologists.

In his *Summa Theologiae* (ca. 1270), Thomas Aquinas laid out a conclusive argument for the existence of God and why life on earth is God given. In *Brief Candle in the Dark*, Dawkins proclaims a "Summa Evolutionis," so to speak: life evolves not by grace of God, but by "selfish genes," a term based on notions of W. D. Hamilton that Dawkins formalized in 1976. Evolution is directed by successful replicators (for example, DNA, sometimes RNA) that influence the development of vehicles (for example, humans, sometimes viruses) that make those vehicles good at reproducing. For almost a century, Darwinists assured us that *individuals* are the targets for natural selection. Dawkins proposes that *genes* are the ones on trial for fitness; individuals serve as their vehicles ("taxi-cabs" in a Japanese translation) to the gyms of selection. He rebukes the late Stephen Jay Gould for not quite getting the replicator–vehicle distinction right: Gould's "genius for getting things wrong matched the eloquence with which he did so." He's also impatient with epigenetics, the mechanics of which he likens to origami. Authors of the approximately 14,000 articles on epigenetics cited by PubMed since 1964 might disagree.

Brief Candle in the Dark presents cogent summaries of Dawkins's other conceptual contributions to biology. One of these made it into the Oxford English Dictionary: the "meme," which Dawkins explains as "a unit of cultural transmission or a unit of *imitation.* Examples of memes are tunes, ideas, catch-phrases, clothes fashions, ways of making pots or of building arches." He's perhaps proudest of the concept of the "extended phenome": species-specific, extracellular structures such as beaver dams, termite towers, coral reefs, and so on (taxi ranks?). Next, we're introduced to the "Concorde fallacy," exemplified by the Anglo-French fiasco in supersonic investment. Dawkins and Jane Brockman use the term in a paper to describe the behavior of digger wasps: "it amounts to investing further in a project simply because one has invested in it heavily in the past." Finally, Dawkins sums up his formulation of "arms races between and within species" as, "the rabbit runs faster than the fox, because the rabbit is running for his life while the fox is only running for his dinner." "Arms Races . . . " is also the title of his most cited scientific paper (1712 citations in PubMed).

Filled with solid science, *Brief Candle in the Dark* is also larded with cameos of life among anglophone elites of science and the mass media. Dawkins's reach is very wide and very cool, moving from high table at Oxford to film festivals in Cannes, from lunch with Claire Bloom to dinner with Bill Gates, from a spat with Ann Coulter to debates with the archbishop of Canterbury. He's amused by a 3,000-year-old [*sic*] dinosaur skull at Liberty University in Lynchburg, Virginia, and he seeks a kraken off the Japanese coast. In the New Mexico desert, a "soft beauty" offers him Ecstasy (the drug); over lunch at Buckingham Palace, the queen of England comments on his hand-painted necktie. Some might consider the inclusion of these anecdotes to be shameless name-dropping, I'd call it "meme-dropping": each conveys a unit of cultural transmission. I also find it refreshing to read a book the index of which places the author's film *The Genius of Charles Darwin* immediately above the names of "George and Ira Gershwin," or which lists the TV comedian "Maher, Bill" right above my favorite poet "Marvell, Andrew."

The book, a constant delight to read, is seasoned with snippets of Dawkins's occasional verse. From 1995 to 2008, he was Simonyi Professor for the Public Understanding of Science at Oxford, and that understanding has been helped by his skill at both exposition and rhyme. Dawkins's verses have clearly helped to form covalent bonds between his friends in the Two Cultures. These lines are in praise of Dawkins's benefactor, the silicon billionaire Charles Simonyi, who owns the Villa Simonyi, a glass ziggurat featuring op art, on the shores of Lake Washington in Medina, Washington:

> Never mind about John Keats
> Or Newton's scientific feats
> Forget your William Butler Yeats
> William Wordsworth, William Gates . . .
> There's the finest champagne and the best from the deli
> (The walls are of glass, when they're not Vasarely)

The God Delusion (2006) was a bestseller that placed Dawkins among the "Four Horsemen of Atheism" along with Christopher Hitchens, Daniel Dennett, and Sam Harris. One dissenting critic, Nathaniel Comfort, called the book an "ecclesiophobic diatribe," and stated, "The gospels of [the Four Horsemen] form the scripture of the 'new atheism,' a fundamentalist sect." Dawkins's Darwinian reproach of intelligent design and/or creationism, beliefs held by a majority of Americans, has earned him a high spot on the opponent list of the Creation Science Hall of Fame. (I'm not displeased to find that the *FASEB Journal* and I have also made it on that roll of honor). In *Brief Candle in the Dark*, Dawkins reviews a decade of jaundiced—and mainly humorless—responses to his earlier work: he's aroused the ire of many and the fury of some. A resolution was even introduced in the Oklahoma legislature to prevent Dawkins from speaking at the University of Oklahoma "to present a biased philosophy of the theory of evolution." But Dawkins prevailed, and describes other amusing encounters with televangelists and other troglodytes of the media on both sides of the Atlantic. To them he responds:

> For what is said in holy writ
> I'm one who doesn't care a bit.
> Away with actuarial mystics!
> I'll throw my lot with hard statistics
> The bible may be old and quaint . . .
> Necess'rily so . . . it aint

Fittingly, at the site of the great Oxford "monkey debate" (1860) between Thomas Henry Huxley and Bishop Wilberforce, Dawkins faced down colleague John Lennox, a strong fan of his deity. With calm, Darwinian reason, Dawkins demolished the creationist cant of his opponent. *Brief Candle in the Dark* clearly establishes Dawkins as the natural heir to Huxley's nineteenth-century mantle as "Darwin's Bulldog" (in Oxbridge parlance, the "bulldog" is the campus cop.) At his seventieth birthday celebration, Dawkins rhymed a wish his many readers would surely support:

[To] reach that bourn—the one we learn
From which no travelers return:
That decent inn—no Marriott—
Presaged by time's winged chariot . . .
Time, yet new rainbows to unweave
Ere going on Eternity leave.

22.

Eugenics and the Immigrant: Rosalyn Yalow

My father, Simon Sussman, was born on the Lower East Side of New York, the Melting Pot for Eastern European immigrants. . . . Since I could type, [I] obtained a part time position as a secretary to Dr. Rudolf Schoenheimer, a leading biochemist at Columbia University's College of Physicians and Surgeons (P&S). This position was supposed to provide an entrée for me into graduate courses, via the backdoor, but I had to agree to take stenography.

—Rosalyn Yalow

On 19 December 1946, Renato Dulbecco and I sailed from Genoa on board the Polish ship, the *Sobieski,* I headed for St. Louis and he for Bloomington. When the Statue of Liberty became visible against the sky of the port of New York . . . I felt as hundreds of thousands of refugees have felt, in the flight from recent as well as earlier persecutions upon arrival in New York Harbor. . . . My stay lasted thirty years.

—Rita Levi-Montalcini

THE BIOGRAPHIES OF ROSALYN YALOW AND RITA LEVI-MONTALCINI yield prima facie arguments for liberal immigration and visa policies. Yalow's story particularly illustrates how bigotry and eugenic notions paradoxically led to scientific discovery—in her case, the discovery of radioimmunoassay—in the United States. Arrivals like those of Yalow's grandfather from czarist Russia or Levi-Montalcini from fascist Italy are perhaps only small episodes in the story of America's rise to pre-eminence in science. Indeed, other factors surely played larger roles: the GI Bill of Rights, James Shannon's NIH, public access to higher education (Hunter College for Yalow), private philanthropy (Rockefeller, Hughes), and more. But we'd surely be many notches down in science had our borders been closed to arrivals from abroad.

Thirty years ago, its repute at apogee, the United States accounted for about 40 percent of the total number of reputable scientific papers published in the world, the European Union for 33 percent, and the Asia-Pacific region for 14 percent.

Those days are over: the seats of American power have been usurped by fans of unreason, Bible-thumpers who feel free to preach "creation science," "alternative medicine," "faith-based" social service, and blatant homophobia. In consequence, the standing of American science has been eroded. By 2004, the EU had moved into the lead with 38 percent of total scientific papers published worldwide and the United States had slipped to 33 percent, while the Asia-Pacific region moved up rapidly to become the source of 25 percent of all papers.

It is not helpful to disguise bans on scientific exchange under the scoundrel's cloak of national security. "Scientists Denied U.S. Visa," a headline screams, and the president of Intel complains to the *Financial Times* that

> America is experiencing a profound immigration crisis but it is not about the 11 million illegal immigrants currently exciting the press and politicians in Washington. The real crisis is that the U.S. is closing its doors to immigrants with degrees in science, maths and engineering.

Data on the U.S. work force in science can be used to make another argument for liberal immigration and visa policies. The 2010 census documented that, whereas Asians are only 6.1 percent of the total work force in the U.S., 17.2 percent of all U.S. life scientists and 19% of all physicians are Asians! We can be grateful that, in deference to our wartime alliance with China, the Roosevelt administration in 1943 repealed the racist Chinese Exclusion Act, which had essentially excluded all Asians from the continental United States since 1881.

I have further reason to thank FDR when I examine publications that crossed my editorial desk even a decade ago. In the March 2006 issue of the *FASEB Journal*, one counts 41 articles (Research Communications and FJ Express) with 312 authors listed, an average of 7.6 per article. Of those authors, 98 (31 percent) had overtly Asian surnames (Indian, Chinese, Japanese, etc.), split evenly between scientists working in U.S. labs and abroad. That squares with the 17 percent of Asians in the work force of life sciences overall in the United States.

In the March 24, 2006, issue of the *Journal of Biological Chemistry*, one could count 65 articles with 399 authors listed, an average of 6.1 authors per article. Of those authors, 188 (47 percent) had Asian surnames (Indian, Chinese, Japanese), with 83 (21 percent) working in U.S. labs and 105 in labs abroad (26 percent). Now let's compare those data with the years of the American apogee.

The March 25, 1986, issue of the *Journal of Biological Chemistry* contained 76 articles with 260 authors listed, a more modest average of 3.4 per article. Of those authors, 44 (16 percent) had Asian surnames, again evenly split between Asians working in U.S. labs and abroad. That's less than half the number of 2006!

That doubling of Asian contributions to American science between 1986 and 2006 is directly due to liberal visa and immigration policies. A generation before, these had brought Renato Dulbecco to Bloomington to work with Salvador Luria on oncoviruses and led Rita Levi-Montalcini to discover nerve growth factor with Stanley Cohen in St. Louis rather than in Turin.

(There were Nobel prizes for all—and for Rosalyn Yalow.)

Rosalyn Yalow's evocation of the Lower East Side of New York as "the Melting Pot for Eastern European immigrants" reminds one that entry of Eastern European Jews into Anglo-Saxon lands was as much of a political issue at the dawn of the twentieth century as is Mexican immigration at the dawn of the twenty-first. In Britain, their exclusion was championed by two leaders of the eugenics movement, Francis Galton and his student Karl Pearson.

In 1925, Karl Pearson, together with Margaret Moul, published an extensive two-part analysis of "The Problem of Alien Immigration into Great Britain, Illustrated by an Examination of Russian and Polish Jewish Children." The paper was the lead article in the *Annals of Eugenics* published by the Galton Institute (K. Pearson, ed.). By means of a detailed study, carried out before World War I, of over a thousand Jewish schoolchildren recently arrived in England from Eastern Europe, the authors attacked the problem of whether the intelligence of these immigrants differed from that of the native stock.

The study was a model of biometric detail: Pearson remains a scientist of repute, whose contributions to biostatistics have remained practically untarnished. Not only were the children he studied given the most modern tests of intelligence, but also school records were examined, home visits made, and physical examinations performed. Control groups were found: English-born Jews and native Gentiles. Elaborate scoring systems were employed to evaluate such variables as size and income of family, rent paid, foci of infection, crowding, ventilation, mouth-breathing versus nose-breathing, "cleanliness," and so on.

The authors directed their inquiry to an applied end, in keeping with the overall aims of the eugenics movement:

> We hold therefore that the problem of the admission of an alien Jewish population into Great Britain turns essentially on the answer that may be given to the question: Is their average intelligence so markedly superior to that of the native Gentile, that it compensates for their physique and habits certainly not being above (probably a good deal below) the average of those characters here?

Pearson and Moul found that, for all groups examined, there was no correlation between intelligence and any other variable such as cleanliness, mode of breathing, family size or income, foci of infection, and height-for-age. Consequently, they were led to this rather somber conclusion:

> . . . the argument of the present paper is that into a crowded country only the superior stocks should be allowed entrance, not the inferior stocks, in the hope—unjustified by any statistical inquiry—that they will rise to the average native level by living in a new atmosphere. The native level is not a product of the atmosphere, but of centuries of racial history, selection, hybridisation and extermination.

Extermination? Be that as it may, the authors failed to note a curious anomaly among their data. All variables considered, there was a striking difference in "intelligence" between Jewish girls and Jewish boys, the latter being statistically more intelligent:

> Namely, that with the Gentile children we have found only a slight difference between the boys and girls. Hence the intelligence of the Jewish girls being much below that of the Jewish boys, even if the latter equaled that of the Gentile boys, the Jewish girls would fall very seriously behind the Gentile girls.

One must point out the genetic fallacy here. If conclusions from such data were possible, we could with some degree of confidence say that in Eastern Jews, by some unusual genetic aberration, intelligence was sex linked, whereas in Gentiles this higher faculty was not.

In the event, arguments such as these directed the bulk of immigrants from Eastern Europe to the Lower East Side of New York rather than to Cheapside in London. We therefore have Karl Pearson and his fellow eugenicists to thank for their indirect gift to American science: permitting Rosalyn Yalow with Solomon Berson to develop radioimmunoassay at a Veterans Administration Hospital in the Bronx.

23.

Cortisone and the Burning Cross

RHEUMATOLOGY, THE TREATMENT OF BONES AND JOINTS and widespread miseries, came late to the game of medical science. For many years my medical specialty was a descriptive art; we had no idea, in any meaningful way, of what was going on. The heart doctors had their cardiograms and digitalis, the endocrine people had their thyroid tests and extracts, but joint doctors seemed condemned to stand idly by and watch their patients turn into cripples after one or another stopgap treatment. Oh yes, we had diathermy, gold salts, paraffin injections, and, believe it or not, bee venom. We knew how to treat gout with colchicine and learned to give penicillin to prevent rheumatic fever, but by and large our treatment of joint disease, or serious threats like systemic lupus erythematosus (SLE), was limited to aspirin, aspirin, and more aspirin. All that changed in 1948, the annus mirabilis of our field. That's the year that cortisone was first given to a patient with arthritis. It's also a year when bigots were burning the houses of black people in white suburbia and lighting crosses on their lawns.

At a staff meeting at the Mayo Clinic in January 1948, Malcolm M. Hargraves described a strange kind of cell that formed in blood samples of patients with SLE. The disease, which tends to afflict young women, attacks joints, skin, kidney, heart, and brain. Before 1950, we couldn't really tell who had SLE and who didn't; we had no clue as to why it was so often fatal. Hargraves had discovered what he called the LE cell, which finally permitted us not only to make a diagnosis of the disease, but also told us what was going wrong with these women. The LE cell, it turned out over the years, is a white blood cell (a neutrophil) that has ingested the dying nucleus of another cell against which lupus patients make antibodies. It also turned out that those antibodies against the nucleus and/or its constituents—the anti-DNA antibodies—were just the tip of an iceberg. SLE patients make a dazzling number of antibodies against bits and pieces of their own cells. Their immune system recognizes such bits of "self" as if they were microbes, tads of "nonself" that want expunging. Hargrave's discovery of the LE cell sparked the study of autoimmunity and lifted rheumatology over the threshold of science.

In the same month, immunologist Harry Rose and rheumatologist Charles Ragan of Columbia described a factor in the serum of most patients with rheumatoid arthritis (RA) that clumped sheep red blood cells coated with human antibodies, the basis of the "sensitized sheep cell agglutination test." Tests for this factor not only permitted accurate diagnosis of rheumatoid arthritis, but also taught us how joints are attacked in RA. What came to be called "rheumatoid factor" turned out to be yet another autoantibody, of great size and with a tendency to form sludge in the blood. Normal human antibodies, the "self" in this case, were recognized as "nonself" by rheumatoid factor. The agglutination reaction in a test tube was a pretty good reflection of what happens in life. In patients with RA, blobs of antibodies containing rheumatoid factor form in the blood like îles flottantes; they become trapped in joint spaces, and joint cells try to get rid of the unwanted debris, cry havoc, and let loose the dogs of inflammation. As with Hargraves and the LE cell, the discovery of rheumatoid factor made it possible to make sense of yet one more of our diseases.

On April 20, 1949, William A. Laurence of the *New York Times* broke news of another discovery announced at a staff meeting at the Mayo Clinic:

> Preliminary tests during the last seven months at the Mayo Clinic with a hormone from the skin of the adrenal glands has opened up an entirely new approach to the treatment of rheumatoid arthritis, the most painful form of arthritis, that cripples millions, it was revealed here tonight.

That evening, Philip Hench, Charles Slocumb, and Howard Polley reported their experience with fourteen cases of rheumatoid arthritis treated with a precious material called "Kendall's compound E" or 17-hydroxy-11-dehydrocorticosterone. Cortisone had entered the clinic.

Within a week, cinemas nationwide showed newsreels of cripples rising miraculously from their wheelchairs. By May 1949, Hench and coworkers reported the "complete remission of acute signs and symptoms of rheumatoid inflammation" at the Association of American Physicians in Atlantic City. In June they added success with rheumatic fever to the cortisone legend at the Seventh International Congress of Rheumatic Diseases in New York. It was the summer I decided to follow my father into rheumatology and, in retrospect, I'd guess that it was cortisone that convinced me. I will never forget the waves of applause after Hench's dramatic film clips were shown to a packed crowd at the Waldorf-Astoria. Now we could actually do something about a crippling disease like rheumatoid arthritis.

In October 1950, the Nobel committee announced that Philip Hench and the two biochemists who had painstakingly isolated and described the chemistry of adrenal steroids, Thaddeus Reichstein (University of Basel) and Edward Kendall (Mayo Clinic), would receive the Nobel Prize in Physiology or Medicine for "their discoveries relating to the hormones of the adrenal cortex, their structure and biological effects." Hench remains the only rheumatologist among Nobel laureates. So universal was the acclaim for cortisone that the Swedish announcement of the 1950 Nobel Prize in Literature, to William Faulkner, was almost a footnote in the world press.

There were other footnotes in the fall of 1950. On Thanksgiving eve, November 22, there was a hate crime in the exclusive Oak Park suburb of Chicago: "Arson Fails at Home of a Negro Scientist" headlined the *New York Times*. It was one of a string of cross-burnings and arson attempts in the white suburbs of Chicago. The scientist in question was Percy Lavon Julian (1899–1975), the first African-American to buy a home in Oak Park. Julian was described in the story as "director of research in the soya products division of the Glidden Company . . . widely acclaimed for his discovery of life stimulating chemicals [and] drugs for treatment of diseases." More to the point, in November 1950 Julian was working feverishly on the practical synthesis of cortisone using Reichstein's compound S, work that resulted in U.S. patent #2,752,339, "Preparation of Cortisone."

On December 11, 1950, less than a month after Julian's house was torched in Oak Park, Hench addressed the Nobel audience at the Karolinska Institutet. He rehearsed the long trail of his discovery: how in the twenties he had first noted relief of rheumatoid arthritis in a male physician who had developed jaundice; how in the thirties he had noted that pregnancy relieved the disease in female patients; how in the forties he had discussed with Kendall the possibility that substance X in the blood of jaundiced or pregnant patients might be his compound E; finally, how in September 1948 he had written to Merck for small amounts of the laboriously synthesized material to test in the clinic. His letter noted that jaundice or pregnancy brought almost immediate relief; he promised Merck that "if any adrenal compound is of real significance in rheumatoid arthritis we would expect to see some results within a very few days." Three days to be exact. Beginning at 100 mg/day, given intramuscularly, the Mayo doctors obtained dramatic results; they soon lowered the dose to a maintenance dose of 25 mg of "cortisone," Hench and Kendall's new name for compound E. That's equivalent to 25 mg tapering to 5 mg of prednisone, and nowadays those results are duplicated daily the world over.

In his Nobel speech, Hench reminded his audience how difficult it was to manufacture practical amounts of cortisone. Merck had got into the steroid business during World War II when the National Research Council subsidized a crash program for synthesis of adrenal steroids. Washington had learned that Luftwaffe doctors were experimenting with injections of adrenal extracts to keep their aviators stress-resistant at 40,000 feet, and several of Kendall's compounds (E and F especially) seemed likely candidates. Merck's Lewis Sarret came up with a complex and difficult synthesis of E from bile: by 1944 it had produced 15 mg from the bile of 2,500 cows! Hench averred that "although none of the thirty-six steps required to convert desoxycholic acid into cortisone has been by-passed, some of the steps have been made less costly, less time consuming, and productive of greater yields." Hench's fellow laureate, Thaddeus Reichstein, told his Stockholm audience, "For practical purposes [the Sarrett] method is much too laborious. In the last two years, again particularly in the U.S.A., at the cost of a considerable amount of time, much better methods have been discovered [among them by] Julian and his collaborators. . . . For after the clinical results of Hench, Kendall and their colleagues it can hardly be doubted that the future demand for these substances will be very great."

"Future demand" was met as the cost of production of cortisone fell from $1000/gm in 1948 to $150 in 1950 to less than $7 in 2000. We owe this boon to the synthesis of cortisone from vegetal sources by Percy Lavon Julian, that brilliant "Negro scientist" whose house in Oak Park was torched on Thanksgiving eve of 1950.

Percy Lavon Julian of Montgomery, Alabama, was the grandson of a former slave and son of a postal employee. He worked his way through DePauw University waiting on tables and graduated as class valedictorian with a Phi Beta Kappa key. After DePauw, he served teaching stints at several black colleges and finally received a fellowship to Harvard, where he earned an MA in chemistry. Since Harvard in the 1920s had no place for a black scientist, Julian applied—successfully—for a Rockefeller Foundation fellowship at the University of Vienna to work with the eminent chemist Ernest Späth. He received his PhD in 1931, having dazzled the Viennese with his skills at tennis and piano, fallen in love with opera, and acquired a long-term collaborator, Josef Pikl.

Julian and Pikl returned to DePauw, taught chemistry, and within four years came up with the total synthesis of physostigmine from the calamar bean (*Physostigma venenosum*). Physostigmine was for many years the only weapon doctors had to fight glaucoma. The bean also contained stigmasterol, an intermediate in sex steroid synthesis, and Julian sought a more abundant source of plant sterols. He wrote to the Glidden Company, a natural-product giant, requesting gallons of soybean oil. This contact led to a job interview at Glidden's labs in Appleton, Wisconsin. But Appleton had a hoary statute on its books dictating that "No Negro should be bedded or boarded in Appleton overnight." Chance favored the prepared chemist, and Julian was offered a far better job in Chicago as director of research of the Soya Products Division of Glidden. The rest is chemical history. In more than eighteen years at Glidden, Julian developed lecithin granules, Glidden's soya oil, and Durkee's edible emulsifiers, and—not incidentally—worked out the commercial syntheses of testosterone and progesterone from soybean oil. He used soy proteins to coat and size paper, to make cold-water paints practical, and to size textiles. During World War II Julian invented AeroFoam, a soy protein product that quenches gasoline and oil fires; the foam saved lives from Europe to the Pacific. Julian was granted more than 100 chemical patents, and Big Pharma still prepares hydrocortisone from compound S by Julian's method. In 1953, he founded his own company, Julian Laboratories, Inc., with labs in the United States and Mexico. In 1961, the company was sold to Smith Kline & French for $2.3 million, "a staggering amount for a Black man at that time."

In his lifetime, Julian was honored by membership in the National Academy of Sciences, a U.S. Postal Service stamp, a dozen honorary degrees, directorships galore, and three public schools that bear his name. We also remember that this agile chemist made it possible to make cortisone from beans instead of bile so doctors could give it to patients for a pittance.

Ave Atque Vale

24.
Lewis Thomas and the Two Cultures

Bien écrire, c'est tout à la fois bien penser, bien sentir et bien rendre; c'est avoir en même temps de l'esprit, de l'âme et du goût. Le style suppose la réunion et l'exercice de toutes les facultés intellectuelles.

[To write well is at once to think, feel and express oneself well; simultaneously to possess wit, soul and taste. Style comes from the integration and exercise of all the intellectual faculties.]

—Comte de Buffon

I remember the moment that Lewis Thomas asked me to be his first chief resident in the old redbrick tenement of Bellevue Hospital. He had a reputation of charming young doctors into academic medicine under conditions and for wages that few dockworkers would tolerate. At the time I was still deciding on whether to follow my father into practice or move on to an academic career. Thomas told me that his chief resident wouldn't have an office, but a lab. He'd be off every third night, and the job would be the first step on the academic ladder. Then Thomas told me what academic salaries were like in the late 1950s, and my rude, younger self quoted: "What is science but the absence of prejudice backed by the presence of money?"

"Henry James," Lewis Thomas snapped, "from *The Golden Bowl*, Chapter One." He went on, "All right then, you won't earn very much, but you'll have a lot of fun in the lab and time to read. If you're lucky, you may also discover something. It's a great life."

The life of Lewis Thomas spanned the golden age of American medicine, an era when—in his words—our oldest art became "the youngest science." Thomas played a major role in that transformation; he was known among scientists as an innovative immunologist, pathologist, and medical educator. He became far better known as a deft writer whose essays bridged the two cultures by turning the news of natural science into serious literature. Witty, urbane, and skeptical, he may have been the only member of the National Academy of Sciences to have won both a National Book Award and an Albert Lasker Award. He is certainly the only medical school dean whose name survives on professorships at Harvard and Cornell, a prize at Rockefeller University, a laboratory at Princeton, and a book that is eleventh on the Modern Library's list of the best 100 nonfiction books of the twentieth century.

Thomas made three important discoveries in immunology, the field of which he was a pioneer; each had implications for human disease. He found that the white cells of blood, the leukocytes, were important mediators of fever and shock brought about by bacterial endotoxins; this taught us how microbes kill us if we don't first kill them. He also made the novel observation that proteolytic enzymes such as papain could injure cartilage when injected into the circulation and the same sort of damage results when our own cells release papain-like ferments; this line of investigation showed us how joints destroy themselves in arthritis. But perhaps his most prescient suggestion, made years before the HIV pandemic, was that our immune system is constantly patrolling our body to find and destroy aberrant cancer-prone cells; we now attribute Kaposi's sarcoma and other AIDS-related tumors to defects in Thomas's "immune surveillance." Those discoveries were made during a very intense period of bench research (1950–1965) at the University of Minnesota and at NYU before he turned his attention to broader issues of science and to his writing.

The lifetime of Lewis Thomas coincided with a special period in American medicine, a time when its scientific base became the strongest it had ever been and its social impact the greatest. Indeed, judging from the numbers who came from overseas to learn from it, American medicine became the envy of the world. That shift of balance from the old world to the new happened at the same time that doctors dropped the laying on of hands and took up monitoring machines. It was not by accident but by design that American medicine was turned from a nineteenth-century folk art into a twentieth-century—and now a twenty-first-century—science. After the Flexner Report of 1910, medical instruction became largely concentrated in university hospitals where the modern sciences of immunology, biochemistry, and genetics could be pursued as eagerly at the bedside as in the lab. Lewis Thomas and his generation of immunologists presided over the conquest of polio and rheumatic fever and the achievements of blood-banking, cardiac surgery, and the transplantation of organs, not to speak of the discovery that DNA was the basic unit of genetic information. In the words of C. P. Snow, they had the future in their bones. Like Snow himself, they were, in the main, committed skeptics.

LEWIS THOMAS GREW UP AS A BRIGHT LAD in a loving family in a comfortable house in Flushing, Queens. His father, Dr. Joseph Simon Thomas (Princeton, 1899; Columbia Physicians and Surgeons, 1904), was a good-natured, hard-working doctor who had met the love of his life, Grace Emma Peck of Beacon Falls, Connecticut, at Roosevelt Hospital, where she was a nurse and he was an intern. They were married on October 30, 1906, and thereafter, in the words of her son, Emma Peck's nursing skills were "devoted almost exclusively to the family."

Lewis Thomas was born on November 25, 1913. As were his three older sisters and younger brother, Lewis was sent to the local schools. But soon the family decided that Flushing High School was not quite ready to prepare another Thomas for Princeton. After three semesters in Queens, Lewis Thomas transferred to the McBurney School, a less than exclusive prep school in Manhattan. He graduated in 1929 in the top quarter of his class. Medical practice was to protect the Thomas family against the worst of the Great Depression, which began on Black Tuesday, exactly one month after fifteen-year-old Lewis left for Princeton in September 1929.

At Princeton he "turned into a moult of dullness and laziness, average or below in the courses requiring real work." He took little interest in physics or inorganic chemistry and dismissed athletics as a general waste of effort. By reason of youth and family standing, he ranked low in the eating club hierarchy of prewar Princeton and was grateful to find safe haven at Key and Seal, a club that was, literally, the farthest out on Prospect Avenue. But high spirits and natural wit brought him to the offices of the *Princeton Tiger*, where Thomas soon published satires, poems, and parodies under the nom de plume "Eltie. "After the crash of '29, we were in thrall to Michael Arlen; we slouched around in Oxford bags and drank bootleg gin from the tub like Scott and Zelda," Thomas recalled. "They told us we'd go out like a light from that stuff. Out like a light. I think I did a piece on bootleg gin for the *Tiger* about that." He had; it's unreadable. Then, on a winter weekend visit to Vassar in 1932, Lewis Thomas met a young freshman from Forest Hills. Her father was a diplomat, her name was Beryl Dawson, and after years of separation for one or another reason they were married a decade later. By then the moult had spread its wings.

Years later, his editor, Elisabeth Sifton, asked me, "When was it that Thomas became so wise?" Thomas attributed his metamorphosis to his senior year at Princeton and a biology course with Professor Wilbur Swingle. Swingle's discovery of a lifesaving adrenal cortical extract—a crude version of deoxycorticosterone—had won wide acclaim. Thomas recalled that Swingle sparked his lifelong interest in the adrenals. Swingle also introduced him to Jacques Loeb's literary/philosophical speculations on ions and cell "irritability" in *The Mechanistic Conception of Life* (1912). Five years out of Princeton, young Thomas would sign up to work with Jacques Loeb's son, Robert F. Loeb, at Columbia.

In his senior year, the depression hit home, and Thomas knew that getting into medical school was one solution to the unemployment problem. He also confessed he had a leg up on other applicants:

> I got into Harvard . . . by luck and also, I suspect, by pull. Hans Zinsser, the professor of bacteriology, had interned with my father at Roosevelt and had admired my mother, and when I went to Boston to be interviewed in the winter of 1933 [Zinsser] informed me that my father and mother were good friends of his, and if I wanted to come to Harvard he would try to help. . . .

Help he did, and Thomas entered Harvard at the age of 19 in the fall of 1933. Thomas's career at Harvard Medical School turned out just fine; he received grades far better than he had at Princeton. When asked in 1983 which member of the Harvard faculty had the greatest influence on his medical education, Thomas replied, "I no longer grope for a name on that distinguished roster. What I remember now, from this distance, is the influence of my classmates." Nevertheless, some on that roster made a lasting impression. Hans Zinsser in bacteriology showed that it was possible to function both as a laboratory scientist and a respected writer; Walter B. Cannon in physiology taught him that the details of homeostasis held the keys to *The Wisdom of the Body*; David Rioch in neuroanatomy had him build a wire and Plasticene model of the brain, with which Thomas trekked about for fifteen years; and in Tracy Mallory's office Thomas came across a pickled specimen that, "like King Charles's head," would haunt his investigative career for decades to come.

At one of Mallory's weekly pathology seminars in the depths of Massachusetts General Hospital, Thomas leaned back in his chair and by accident knocked over a sealed glass jar containing the kidneys of a woman who had died of eclampsia. Replacing the jug, he noted that both organs were symmetrically scarred by the deep, black telltale marks of bilateral renal cortical necrosis. Thomas remembered having seen something like those pockmarked kidneys before. They had been provoked in rabbits by two appropriately spaced intravenous injections of endotoxin: the effect was called the generalized Shwartzman phenomenon, and he would tussle with it for the rest of his scientific career.

Thomas graduated cum laude from Harvard Medical School in 1937 and began internship on the Harvard Medical Service of Boston City Hospital. A history of the Harvard Medical Unit at Boston City Hospital documents that of the 71 young physicians who trained there between 1936 and 1940, 52 became professors of medicine, and 6 went on to the deanship of medical schools.

Thomas remained at Boston City until 1939, when the confluence of his interests in neurology, adrenal hormones, and the Loeb mystique brought him to New York. Halfway through his internship in Boston, he heard that Dr. Robert F. Loeb was becoming director of the Neurological Institute in New York, and he resolved to study with him because

> Loeb was a youngish and already famous member of the medical faculty in the Department of Medicine at P&S, recognized internationally for his work on Addison's disease [and] the metabolic functions of the adrenal cortex and the new field of salt and water control in physiology.

Thomas served as a neurology resident (his only specialty training) and research fellow at P&S from 1939 to 1941, with time out to marry Beryl at Grace Church in New York in January 1941. Robert Loeb abruptly moved to the chairmanship of the Department of Medicine, but Thomas found that there was a fellowship with John Dingle awaiting him back at Harvard, and he jumped at the chance.

Almost as soon as Lew and Beryl had established themselves back in Boston, Thomas was sent by Dingle on a month-long medical mission to Halifax, Nova Scotia, where an outbreak of meningococcal meningitis had struck the wartime port. Beryl served as lab assistant. Those four weeks in the field, an important publication on the effects of sulfadiazine in meningitis, and a thorough grounding in immunology in Dingle's lab were a prelude to a naval commission after Pearl Harbor. Thomas reported in March 1942 to the Naval Research Unit at the hospital of the Rockefeller Institute in New York and on January 12, 1945, landed with a detachment of the unit headquartered in Guam. Thomas and Horace Hodes were put to work on Japanese B encephalitis on Okinawa, and quickly identified horse blood as a reservoir for the virus.

War ended; waiting to be sent stateside, Thomas began experiments on rheumatic fever in Guam. Putting unused lab facilities to good purpose, he knew that rheumatic fever was almost always preceded by a streptococcal throat infection and that the interval between infection and heart disease could be very long indeed. Perhaps the disease was an allergic reaction to the microbe, to the patient's own tissues, or to a mixture of the two.

In Guam, Thomas found that rabbits receiving a mixture of microbes (streptococci) and ground-up heart tissue became ill and died within two weeks; the microscope revealed that their hearts showed lesions that resembled those of rheumatic fever in humans. Control rabbits injected with streptococci alone or with ground-up heart tissue alone remained healthy and showed no such damage. Thomas was entirely confident that he had solved the whole problem of rheumatic fever. He hadn't. On his return to the Rockefeller Institute, he couldn't repeat those experiments, sacrificing "hundreds of rabbits, varying the dose of streptococci and heart tissue in every way possible." He was vastly relieved that he hadn't rushed into print on the basis of those rabbits in Guam.

Thomas's first faculty position after discharge from the navy in 1946 was as assistant professor of pediatrics at the Harriet Lane Home for Invalid Children at Johns Hopkins. Thomas tried once more to repeat those rabbit experiments. This time, he mixed streptococci and heart tissue with a devilish brew of fats (the solution called Freund's adjuvant) that had produced tissue injury in other disease models. Bad news for assistant professor Thomas: the rheumatic fever experiments failed once again. But Thomas could not shake off those experiments that had worked so well in Guam. Perhaps the host, the rabbits in Guam, for example, but not those in New York or Baltimore, had been "prepared" by an earlier insult such as the endotoxin injection that provoked the Shwartzman phenomenon.

He tackled the problem with Chandler (Al) Stetson, a lifelong friend who was to become his colleague in Minnesota and his successor as the professor of pathology at NYU. Thomas and Stetson "prepared" rabbits with endotoxin from meningococci. The prepared skin had an excess of acid, and they reasoned that the acid might activate the tissue's own ferments, proteases called cathepsins. But they were neither able to measure cathepsin activity nor obtain purified cathepsins, so they next injected rabbits with enzymes obtained off the shelf, trypsin and papain. Trypsin was ineffective, but papain produced lesions in the skin that looked very much like the local Shwartzman reaction.

When Thomas left Hopkins, he took the problem with him. He served a brief stint at Tulane, where he became a professor of medicine and director of the Division of Infectious Disease. He was diverted for a while by studies of circulating antibodies in animal models of multiple sclerosis but returned to rheumatic fever when he was appointed American Legion Professor of

Pediatrics and Medicine at the University of Minnesota in 1951. In quick time, he put together a team of young investigators, most of whom were soon at work on the Shwartzman phenomenon and the streptococcus.

He reverted to the notion that proteases, either secreted by the streptococcal microbe or released from the victim's own cells, caused damage in a "prepared" heart or joint. With a young Minnesota pediatrician, Robert A. Good ("the smartest investigator I ever met," he once told me), he found out that if one removed white cells from the Shwartzman equation, kidney injury was prevented. The kidneys were also spared if one gave heparin, which prevented blood vessels from becoming plugged by fibrin, platelets, and white cells. Good and Thomas suggested that "a combination of humoral and cellular factors made by the host caused the tissue injury." Nowadays we invoke complicated systems with complicated names such as anaphylatoxins, Toll receptors, apoptosis, caspases, and cytokines to explain the Shwartzman phenomenon. But in the 1950s Good and Thomas had provided a satisfactory explanation, and the flow of satisfying, explanatory papers followed Thomas from Minnesota as he moved to NYU in 1954.

Thomas was recruited to NYU by Colin McLeod to become professor and chairman of the Department of Pathology. He was delighted to return to the metropolis with Beryl and his three daughters, Abigail (born 1941), Judith (born 1944), and Elizabeth (born 1948), and to set up his household at Sneden's Landing, a small town up the Hudson from the city. He remained at NYU for fifteen years and proceeded to turn it into a world center of immunology, first in the Department of Pathology (1954–1958), then as professor and chairman of the Department of Medicine, and finally as dean of the New York University School of Medicine and deputy director of NYU Medical Center (1966–1969).

Over those years he attracted and/or trained a legion of scientific stars and superstars at NYU: Frederick Becker, Baruj Benacerraf, John David, Edward Franklin, Emil Gottschlich, Howard Green, H. Sherwood Lawrence, Robert T. McCluskey, Peter Miescher, Victor and Ruth Nussenzweig, Zoltan Ovary, Jeanette Thorbecke, Stuart Schlossman, Chandler Stetson, Jonathan Uhr, and Dorothea Zucker-Franklin. Thomas's international colleagues were frequent visitors: Sir Macfarlane Burnett, Dame

Honor Fell, Philip Gell, James Gowans, Sir Peter Medawar, Thomas Sterzl, and Guy Voisin.

Early in his NYU days Thomas hit a rough patch. Whereas cortisone, the miracle drug, clearly stopped inflammation in the clinic, Thomas was astonished to find that cortisone not only proved ineffective against the Shwartzman phenomenon but actually provoked it. This puzzle took the wind out of his sails. He was indeed "in irons on his other experiments" and "not being brilliant." Then came the "floppy-eared bunnies": injections of a protease, papain, into New Zealand white rabbits caused their ears to droop as their cartilage melted away.

After papain, new discoveries proceeded apace. If an exogenous protease caused connective tissue damage, where might endogenous proteases reside? Thomas spent a summer with Dame Honor Fell, director of the Strangeways Research Laboratory in Cambridge. Fell had been studying vitamin A and had found that it produced depletion of cartilage matrix in mouse bone rudiments growing in a dish. Fell and Thomas decided to trade experimental systems. They first added papain to the little bone cultures in the dish and were able to produce vitamin A–like lesions in mouse cartilage. Thomas then returned to NYU to do the reciprocal experiment. Together with Jack Potter and R. T. McCluskey, Thomas and I stoked rabbits full of the vitamin A, and sure enough: twenty-four to forty-eight hours later, their ears drooped as if they had been given papain. We were convinced then that vitamin A in some fashion released an endogenous papain-like enzyme from cartilage cells and that this enzyme proceeded to break down cartilage matrix. At the time we supposed that the enzyme was present in lysosomes, subcellular "suicide sacs" that had just been described by Christian de Duve at Louvain. We suggested that vitamin A had ruptured the walls around these organelles, and that cortisone and its analogues must therefore stabilize the lysosomes.

These days the answer is more complicated. We now believe that metalloproteinases are released from cells and that synthesis of these proteases is under opposing transcriptional control by vitamin A and cortisone acting via well-defined cytoplasmic and nuclear receptors. It all seemed simpler a generation ago. But these experiments, the last in which Thomas played a hands-on role, pointed the way for Thomas's students and, in turn, their students, to explore other areas of human biology: how infection and immunity make our white cells clump and stick to blood vessels; how stimulated white cells release molecules (cytokines) that cause fever, fatigue, and inflammation; how tissues are recognized as foreign and transplants are rejected; how cortisone and aspirin-like drugs work in arthritis—and, as a follow-up of the cortisone/lysosome experiments, how to design, manufacture, and bring to the clinic tiny drug-bearing lipid structures called liposomes, which have saved thousands of lives.

After 1965, Thomas moved from the lab bench to the rougher terrain of medical administration and science policy. Thomas had a broad interest in how medical science shapes, and is shaped by, society. Wit, candor, and attention to principle rather than politics made him a valuable spokesman for medical science. While still at NYU, Thomas served as a member of the New York City Board of Health (1957–1969), was instrumental in the construction of the new Bellevue Hospital, and set up the Health Research Council, a sort of local NIH. As chairman of the Narcotics Advisory Committee of the New York City Health Research Council, he guided Vincent P. Dole into methadone research and pointed Eric Simon to endorphins (1961–1963). After a stint in New Haven as professor of pathology and dean (1969–1973) at Yale University School of Medicine, he became president and chief executive officer of the Memorial Sloan Kettering Cancer Center (1973–1980). At MSKCC, he launched a major attack on tumor immunology, recruiting Robert Good as director; Thomas became chancellor of MSKCC from 1980 to 1983. In retirement, his summer home in the Hamptons made a university professorship at SUNY–Stony Brook (1984) convenient, and his Manhattan apartment let him serve as writer in residence at Cornell University Medical School until his death in 1993. By then he had been for several decades the most widely read interlocutor between the older literary culture and the new world of medical science.

Lewis Thomas was preceded in his role as medical scribe to the nation by such other American physician-writers as Oliver Wendell Holmes, William James, Walter B. Cannon, and Hans Zinsser. Each contributed vastly to the biology of medicine, each wrote books that gained a broad general audience, each taught that science is a very human activity. Zinsser, who, as we've seen, was responsible for Thomas's admission to Harvard Medical School, summed up the tradition by proposing that

> Aside from the habits of hard work that [medicine] demands, it embraces a broad survey of the biological field, enforces a considered correlation of the fundamental sciences, and, on the human side, brings the thoughtful student face to face with the emotional struggles, the misery, courage and cowardice of his fellow creatures.

Like Zinsser before him, Thomas fused the two cultures of "fundamental science" and "the human side," because for him they were one. He went on to become a persuasive spokesman for the biological revolution because he was convinced that while science is only one of the many ways we have of making sense of nature, medical science is the only way we have of making sense of disease. That conviction would have remained more or less private were it not for the public language in which it was voiced. His sound was distinct and unmistakable, the prose direct and limpid. Here, for example, is Thomas's suggestion for signals we might send from earth to announce ourselves to whatever life there might be in outer space:

> Perhaps the safest thing to do at the outset, if technology permits, is to send music. This language may be the best we have for explaining what we are like to others in space, with least ambiguity. I would vote for Bach, all of Bach, streamed out into space, over and over again. We would be bragging, of course, but it is surely excusable for us to put the best possible face on at the beginning of such an acquaintance. We can tell the harder truths later. . . . Perhaps, if technology can be adapted to it, we should send some paintings. Nothing would better describe what this place is like, to an outsider, than the Cézanne demonstrations that an apple is really part fruit, part earth.

That sort of writing is the product of another unique period in American culture. Thomas and his colleagues were educated in colleges at which the liberal arts were still firmly in place and John Dewey's learning-by-doing had moved from primary schools into the universities. It was an era when those who did medical science were expected to know why it was done and for whom. They were also expected to make only modest claims for their success: "I was lucky," Thomas quipped after he received a medal at Bologna in 1978, "chance favored the prepared grind." One knew that he was speaking for a generation of medical scientists who believed that one could do serious work without taking oneself too seriously.

Thomas wrote one, rather impersonal, memoir, *The Youngest Science.* He believed that a scientific memoir ought to remember not only how the science came about but how it felt at the time: sometimes right, sometimes wrong, always surprising. Thomas called it a story of finding a pattern in a jumble, a task even he found daunting:

> It should be easier, certainly shorter work to compose a memoir than an autobiography, and surely it is easier to sit and listen to the one than to the other. . . . In my [memoir] I find most of what I've got left are not memories of my own experience, but mainly the remembrance of other people's thoughts, things I've read or been told, metamemories. A surprising number of them turn out to be wishes rather than recollections . . . hankerings that the one thing leading to another had a direction of some kind, and a hope for a pattern from the jumble—an epiphany out of entropy.

That passage, precise and informal at once, illustrates the flow of Thomas's thought and speech. Thomas was as likely on the wards as in print to pair epiphany (as in James Joyce) with entropy (as in the second law of thermodynamics). That balance of phrase could be said to be the signature of Thomas's prose; epiphany seemed to be having it out with entropy on every page. He was also sparing of words when fewer spoke louder. Half a year before his death from Waldenström's macroglobulinemia, a slow, wasting form of bone marrow disease, he received an award named in his honor at

Rockefeller University. Confined to a wheelchair by his illness, he declined the podium and apologized to the audience for "not rising to the occasion." About the same time, I reached him on the telephone:

"How are you doing?" I asked. He knew what I was asking.

"So," he replied.

"What do you mean by 'so'?"

"Well," said Thomas, "in my family, there were only three ways of answering that question of yours. If things were going along splendidly, you'd answer 'fine.' If there were a bit of trouble around, you'd say 'so-so.' Right now, I'm 'so.'"

When more words were required, they flowed like wine. Thomas's chosen means of expression was the informal essay, a literary form that accommodates many topics but always has the mind of its author as the subject. A reader of his pieces quickly becomes aware that Thomas has invited him to a tug of war between two turns of mind, a playful match between two equally attractive personae: a cheerful Thomas who urged holism and a doubting Thomas who was a card-carrying reductionist. Holism, as I learned from Thomas, was invented by General Jan Smuts (the intelligent design chap) in 1926. It implies that matter and life are one. Reductionism derives from Hippolyte Taine, who borrowed the term from the chemists who use it to denote an agent that reduces a compound to a simpler substance by removing oxygen. Matter without life, one might say.

As a medical scientist, Thomas was persuaded that only patience, doubt, and diligence, the reductionist virtues, could pluck facts from nature. But Thomas also understood the very human need to turn the strands of fact into a fabric of belief. In that mode he had but one exemplar: William James. Especially in his later, more ruminative essays Thomas successfully blended the Jamesian "Will to Believe" with James Lovelock's newer Gaia hypothesis, a postulate that life on our planet has been chiefly responsible for the regulation of that life's own environment. Lovelock's holistic notion seems to unite the best of James with the best of Thomas, and it is no mean compliment to suggest that a passage such as this from James would fit comfortably in any of Thomas's essays:

> We find ourselves believing, we hardly know how or why. We all of us believe in molecules and the conservation of energy, in democracy and necessary progress . . . all for no reason worthy of the name. We see into these matters with not more inner clearness, and probably with much less, than any disbeliever in them might possess. . . . For us not insight, but the prestige of the opinions, is what makes the spark shoot from them, that lights up our sleeping magazines of faith. Our faith is faith in someone else's faith, and in the greatest matters this is most the case. Our belief in truth itself is that there is a truth and that our minds and it are made for each other.

Like James, Thomas was celebrated as a "poetic" or "creative" writer and scientist. But modern critics use those adjectives in much the way that eighteenth-century essayists would have used "sentimental." Thomas was by no means a sentimental essayist. I'm convinced that there was more structure than sentiment to his writing, just as there was more science to his art than art to his science. Here is a Thomas passage with a Jamesian sense of our planet:

> The overwhelming astonishment, the queerest structure we know about so far in the whole universe, the greatest of all cosmological scientific puzzles, confounding all our efforts to comprehend it is the earth. We are only now beginning to appreciate how strange and splendid it is, how it catches the breath, the loveliest object afloat around the sun, enclosed in its own blue bubble of atmosphere, manufacturing and breathing its own oxygen, fixing its own nitrogen from the air into its own soil, generating its own weather at the surface of its own rain forests, constructing its own carapace from living parts: chalk cliffs, coral reefs, old fossils from earlier forms of life now covered by layers of new life meshed together around the globe, Troy upon Troy.

Again like James, Thomas was not simply a clever scientist with a creative turn of mind; he was a writer to the bone. Evelyn Waugh, whom

Thomas had admired since his undergraduate days on the *Princeton Tiger*, introduced the term "architectural" to describe the difference:

> Creative is an invidious term [for a writer] . . . a better word, except that it would always involve explanation, would be "architectural." I believe that what makes a writer, as distinct from a clever and cultured man who can write, is an added energy and breadth of vision which enables him to conceive and complete a structure.

The architectural structure that Thomas worked out fitted readily into the conventions of the informal essay. He derived from the facts of natural science, such as the workings of inflammation, symbiosis, the planetary ecosystem, and the life of social insects, metaphors for broad aspects of human activity, such as curiosity, language, and altruism. Fact marched hand in hand with solace; he assured us that a meningococcus with the bad luck to catch a human was in more trouble than a human who catches a meningococcus. Who on earth would not welcome those tidings of comfort and joy? Every once in a while Thomas reversed the direction of his metaphors, using human behavior or language as a metaphor for the odd fact of cell biology:

> The meaning of these stories [of protozoan symbiosis] may be basically the same as the meaning of a medieval bestiary. There is a tendency for living things to join up, establish linkages, live inside each other, return to earlier arrangements, get along, wherever possible. This is the way of the world.

But his years in the lab served him well on the page. His sense of trial and error at the bench and in the clinic, of how cells divide, microbes hurt, and creatures die, gave a tough edge to his writing:

> When injected into the bloodstream, [endotoxin] conveys propaganda, announcing that typhoid bacilli (or other related bacteria) are on the scene and a number of defense mechanisms

> are automatically switched on, all at once . . . including fever, malaise, hemorrhage, shock, coma and death. It is something like an explosion in a munitions factory.

Thomas had been writing for publication since his efforts on the *Princeton Tiger*. His poetry became more ambitious, and in his house office days he published several remarkably polished verses in literary magazines such as the *Atlantic Monthly*. One of these even appeared in 1944 while he was in the Pacific with the Rockefeller Institute Medical Research Unit. But, for twenty years after he returned from the war, his writing was pretty much limited to the scientific literature. Later, after he had made his contributions to immunology, after he had secured his reputation in science, and while he was serving as chairman, dean, and chancellor at three sometimes-exasperating institutions, he turned his attention once more to the muse.

In 1965, he had permitted himself a ruminative essay on inflammation, which was brought to the attention of Franz Ingelfinger, editor of the *New England Journal of Medicine* and an old colleague of Thomas's at Harvard. At Ingelfinger's request, Thomas began his stint as author of the bimonthly column "Notes of a Biology Watcher." Thanks to Elisabeth Sifton, then an editor at Viking Press, those sparkling essays were soon collected into *The Lives of a Cell*, the volume became a best seller, won a National Book Award—and Thomas was well on his way to a place in the world of letters.

Unlike other scientist-writers who tend to limit their subjects to their own field of research—evolution is the core of Stephen Jay Gould, for example—Thomas worked hard at the task of writing well about all manner of things. Waugh would have appreciated that effort:

> In youth high spirits can carry one over a book or two. The world is full of discoveries that demand expression. Later a writer must face the choice of becoming an artist or prophet. He can shut himself up at his desk and selfishly seek pleasure in the perfecting of his own skill or he can pace about, dictating dooms and exhortations on the topics of the day. The recluse at

> the desk has a bare chance of giving abiding pleasure to others; the publicist has none at all.

Tough critics of his science claimed that much of Thomas's immunology was small scale and anecdotal. But Thomas's lasting contributions to our understanding of inflammation and immunity can be readily identified today; his notions are embedded in the history of immunology. On the literary side, Stephen Jay Gould accused Thomas's essays of disguising their serious themes as "homegrown Yankee wisdom" cloaked in "charming and superficially rambling" accounts—too charming, perhaps, for words. Sidney Hook complained about his memoir that "Mr. Thomas's 'A Long Line of Cells' is an instructive lesson in biology but tells us nothing about his life that distinguishes it from any other human life." Christopher Lehmann-Haupt accused Thomas of "Optimism (relentless) on humanity." Thomas himself confessed that he may have told us once too often that all is for the best in this best of all possible worlds: "I'm not sure Pangloss was all that wrongheaded. This is in real life the best of all possible worlds, provided you give italics to that word possible." But his mandarin wit prevailed over the Panglossian strain; he became a fine essayist by the best means available, which was to work hard at writing well. Indeed, his very last, somewhat slight, book, *Et Cetera, Et Cetera,* was devoted entirely to words and the sound of words. That attention to the *mot juste* reminds one of Waugh's advice that over time the writer is better off perfecting his style rather than peddling his subject:

> Literature is the right use of language irrespective of the subject or reason of utterance. A political speech may be, and often is, literature. A sonnet to the moon may be, and often is, trash. Style is what distinguishes literature from trash. . . . The necessary elements of style are lucidity, elegance, individuality; these three qualities combine to form a preservative which ensures the nearest approximation to permanence in the fugitive art of letters.

That point taken, I'd make the partisan argument that Thomas's careful attention to style, which so clearly meets Waugh's criteria of lucidity, elegance, and individuality, gives Lewis Thomas a shot at permanence in the world of letters. A number of his compositions stand up to essays by such other modern masters of the genre as E. B. White, A. J. Liebling, and John Updike. In a select few of his essays, Thomas reaches back to touch the mantle of Montaigne.

Notes

Prefatory Note

Mencken, H. L. *A Mencken Chrestomathy: His Own Selection of His Choicest Writings*. New York: Vintage, 1982. First published 1949.

Thomas, Lewis. *The Lives of a Cell: Notes of a Biology Watcher*. New York: Viking, 1974.

Going Viral

1. *Arrowsmith* and CRISPR at the Marine Biological Laboratory

Baltimore, D., et al. "A prudent path forward for genomic engineering and germline gene modification." *Science* 348 (2015): 36–38.

Chicago Daily Tribune. "Loeb Tells of Artificial Life." December 28, 1900.

Doudna, J. A. "From Bacterial Adaptive Immunity to the Future of Genome Engineering." Glassman Lecture at the University of Chicago, July 1, 2016.

Jinek, M., et al. "A programmable dual-RNA-guided DNA endonuclease in adaptive bacterial immunity." *Science* 337 (2012): 816–821.

Kim Kardashian's Twitter page. https://twitter.com/kimkardashian?lang=en.

Lander, E. S. "The Heroes of CRISPR." *Cell* 164 (2016): 18–28.

Lewis, Sinclair. *Arrowsmith*. London: Jonathan Cape, 1925.

Loeb, Jacques. "On the Limits of Divisibility of Living Matter." In *Biological Lectures at MBL Wood's Holl*. Boston: Ginn & Co, 1894.

Loeb, Jacques. "Experimental Studies on the Influence of Environment on Animals." In *Darwin and Modern Science: Essays in Commemoration of the Centenary of the Birth of Charles Darwin and of the Fiftieth Anniversary of the Publication of "The Origin of Species."* Edited by A. C. Seward. Cambridge: Cambridge University Press, 1909.

Loeb, Jacques. *The Mechanistic Conception of Life*. Chicago: University of Chicago Press, 1912.

Pourcel, C., G. Salvignol, and G. Vergnaud. "CRISPR elements in *Yersinia pestis* acquire new repeats by preferential uptake of bacteriophage DNA, and provide additional tools for evolutionary studies." *Microbiology* 151 (2005): 653–63.

San Francisco Call. "Believes Germ of Life Will Be Discovered." October 1, 1907.

Thomas, L. "The MBL." *New England Journal of Medicine* 286 (1972): 1254–6.

Trilling, Lionel. *The Liberal Imagination*. New York: NYR Books, 2008. First published 1950.

2. Ebola and the Cabinet of Dr. Proust

Ahearn, I. M., et al. "Regulating the regulator: post-translational modification of RAS." *Nature Reviews Molecular Cell Biology* 13 (2012): 39–51.

Aleksandrowicz, P., et al. "Ebola virus enters host cells by macropinocytosis and clathrin-mediated endocytosis." *Journal of Infectious Diseases* 204 (Suppl. 3) (2011): S957–S967.

Annual Report of the Surgeon General, Marine Hospital Service, U.S. Public Health Service, 1892, p. 47.

Brown, DaNeen L. "Texas Officials Transport Ebola Victim for Cremation." *Washington Post*, October 9, 2014. http://www.washingtonpost.com/news/post-nation/wp/2014/10/09/texas-officials-transport-ebola-victim-for-cremation/.

Centers for Disease Control and Prevention. "Enhanced Ebola Screening to Start at Five U.S. Airports and New Tracking Program for all People Entering U.S. from Ebola-affected Countries." News release, October 8, 2014. http://www.cdc.gov/media/releases/2014/p1008-ebola-screening.html.

Central Intelligence Agency. "The World Factbook: Infant Mortality Rate." https://www.cia.gov/library/publications/the-world-factbook/rankorder/2091rank.html.

Cooper, Helene. "Panic Where Ebola Risk Is Tiny; Stoicism Where It's Real." *New York Times*, October 20, 2014.

The Economist. "The Ebola Crisis: Much Worse to Come." October 18, 2014.

Harper, C., et al. "Targeting membrane trafficking in infection prophylaxis: dynamin inhibitors." *Trends in Cell Biology* 23 (2013): 90–101.

Howard-Jones, Norman. *The Scientific Background of the International Sanitary Conferences: The 1866 Conference*. Geneva: World Health Organization, 1975.

Kuhnke, LaVerne. *Lives at Risk: Public Health in Nineteenth-Century Egypt*. Berkeley: University of California Press, 1990.

Macia, E., et al. "Dynasore, a cell-permeable inhibitor of dynamin." *Developmental Cell* 10 (2006): 839–850.

Masaike, Y., et al. "Identification of dynamin-2-mediated endocytosis as a new target of osteoporosis drugs, bisphosphonates." *Molecular Pharmacology* 77 (2010): 262–269.

McCarthy, Mary. *The Group*. New York: Harcourt, Brace & World, 1963.

Mouawad, Jad. "Experts Oppose Ebola Travel Ban, Saying It Would Cut Off Worst-Hit Countries." *New York Times*, October 17, 2014.

NBC News. "Obama Says Ebola Travel Bans Could Make Things Worse." October 17, 2014. http://www.nbcnews.com/storyline/ebola-virus-outbreak/obama-says-ebola-travel-bans-could-make-things-worse-n228731.

New York Times. "Government Officers Anxious; The Power of the Federal Authorities in Quarantine Matters." September 1, 1892.

Oestereich, L., et al. "Successful treatment of advanced ebola virus infection with T-705 (inhibits RNA polymerase) (favipiravir) in a small animal model." *Antiviral Research* 105 (2014): 17–21.

Perez, C., et al. "The Ebola Crisis: Sick Nurse an 'Err Carrier.' CDC Let Her on Jet Despite Fever. Search Is On for 132 Passengers." *New York Post*, October 16, 2014.

Proust, Adrien. *Essai sur l'hygiène international: ses applications contre la peste, la fièvre jaune et le choléra asiatique* ["Essay on International Hygiene"]. Edited by G. Masson. Paris: Librairie de l'Académie de Médecine, 1873.

Proust, Adrien. *La défense de l'Europe contre le choléra*. Edited by G. Masson. Paris: Libraire de l'Académie de Médecine,1892.

Proust, Marcel. *Remembrance of Things Past*. Vol. 2. New York, Random House, 1941. First published 1919.

Robb, G. "Plots in Paradise" (review of J. Ring, *Riviera*). *Daily Telegraph*, June 13, 2004.

Shpetner, H. S., and R. B. Vallee. "Identification of dynamin, a novel mechanochemical enzyme that mediates interactions between microtubules." *Cell* 59 (1989): 421–432.

Weissmann, Gerald. "Cholera at the Harvey." In *The Woods Hole Cantata*. New York: Dodd Mead, 1985.

Weissmann, Gerald. "Ebola: Out of Africa with the Sanitarians." In *The Year of the Genome*. New York: Times Books, 2000.

World Health Organization. "Ebola Situation Reports." March 2016. http://www.who.int/csr/disease/ebola/situation-reports/archive/en/

3. Zika, Kale, and Calligraphy: Ricky Jay and Matthias Buchinger

Barrangou, R. "CRISPR provides acquired resistance against viruses in prokaryotes." *Science* 315 (2007): 1709–1712.

Fell, H. B., and E. Mellanby. "Effects of hypervitaminosis A on foetal mouse bones cultivated in vitro." *British Medical Journal* 2 (1950): 535–539.

Geoffroy Sainte-Hilaire, Isidore. *Histoire générale et particulière des anomalies de l'organisation chez l'homme et les animaux*. Paris: J.-B. Baillière, 1836.

Jay, Ricky. *Learned Pigs and Fireproof Women*. New York: Farrar, Strauss & Giroux, 1986.

Jay, Ricky. *Celebrations of Curious Characters*. San Francisco: McSweeney's Books, 2011.

Jay, Ricky. *Matthias Buchinger: "The Greatest German Living."* Los Angeles: Siglio Press, 2016. Published in conjunction with the exhibition "Wordplay: Matthias Buchinger's Inventive Drawings from the Collection of Ricky Jay," shown at the Metropolitan Museum of Art.

Kelsey, F. O. "Drug embryopathy: the thalidomide story." *Maryland State Medical Journal* 12 (1963): 594–7.

Krautkanal (blog). "Tetraphokomelie." http://krautkanal.com/b/10085800.

McGrath, Charles. "Ricky Jay and the Met Conjure Big Magic in Miniature. *New York Times,* January 13, 2016.

Novotny, J. A., et al. "Plasma appearance of labeled beta-carotene, lutein, and retinol in humans after consumption of isotopically labeled kale." *Journal of Lipid Research* 46 (2005): 1896–903.

Sheikh, B. N., et al. "Excessive versus physiologically relevant levels of retinoic acid in embryonic stem cell differentiation." *Stem Cells* 32 (2014): 1451–8.

Teratology Society. Brazilian Scientists Head to San Antonio, Set to Unveil New Zika Research. June 20, 2016. http://www.prweb.com/releases/new_zika_research/2016/prweb13497384.htm

Vargesson, N. "Thalidomide-induced teratogenesis: history and mechanisms." *Birth Defects Research* 105 (2015): 140–156.

Weissmann, G. "Changes in connective tissue and intestine caused by vitamin A in amphia and their acceleration by hydrocortisone." *Journal of Experimental Medicine* 114 (1961): 581–92.

Wikipedia. https://en.wikipedia.org/wiki/Thalidomide#/media/File:Niko_von_Glasow.jpg

4. Ike on Orlando: "Every Gun is a Theft"

Alpers, Philip, Amélie Rossetti, and Daniel Salinas. "United States—Gun Facts, Figures and the Law." Sydney School of Public Health, The University of Sydney. GunPolicy.org, December 2, 2016. Accessed June 26, 2017. http://www.gunpolicy.org/firearms/region/united-states.

American Battle Monuments Commission. Korean War Casualties. http://www.abmc.gov/search/koreanwar.php

Barnes, Brooks. "'Lego Movie' Is on Top at Box Office." *New York Times*, February 17, 2014.

Berman, Jillian. "Taxpayers Shoulder Bulk of Gun Violence Health Care Costs: Study." *Huffington Post*, September 13, 2013. http://www.huffingtonpost.com/2013/09/13/taxpayers-gun-violence_n_3915434.html?view=print&comm_ref=false.

Bowling, Melissa, and James Moske. "In the Footsteps of the Monuments Men: Traces from the Archives at the Metropolitan Museum." January 31, 2014. http://www.metmuseum.org/about-the-museum/now-at-the-met/2014/in-the-footsteps-of-the-monuments-men.

Brown University. "Estimated cost of post-9/11 wars: 225,000 lives, up to $4 trillion." June 29, 2011. https://news.brown.edu/articles/2011/06/warcosts.

Chrysler. "Chrysler Presidents Day Event TV Commercial, 'Dwight D. Eisenhower.'" February 17, 2014. http://www.ispot.tv/ad/7B7q/chrysler-presidents-day-event-dwight-d-eisenhower

Chumley, Cheryl K. "Ted Nugent, Pre-campaign Trail: Obama's a 'Subhuman Mongrel.'" *The Washington Times*, February 19, 2014.

Daggett, Stephen. "Costs of Major U.S. Wars." Congressional Research Service Report for Congress. June 29, 2010. https://fas.org/sgp/crs/natsec/RS22926.pdf.

Dargis, Manohla. "A-Team Tracks Nazi Plunder: Clooney and Company Hunt for Treasures in 'Monuments Men.'" *New York Times*, February 6, 2014.

DePastino, Todd. *Bill Mauldin: A Life Up Front*. New York: W. W. Norton, 2008.

Edsel, Robert M., with Brett Witter. *The Monuments Men: Allied Heroes, Nazi Thieves, and the Greatest Treasure Hunt in History*. New York: Center Street/Hachette, 2010.

Eisenhower, Dwight D. "Annual Message to the Congress on the State of the Union, February 2, 1953." http://www.presidency.ucsb.edu/ws/?pid=9829.

Eisenhower, Dwight D. "Citation Presented to Dr. Jonas E. Salk and Accompanying Remarks." April 22, 1955. Online by Gerhard Peters and John T. Woolley, *The American Presidency Project*. http://www.presidency.ucsb.edu/ws/?pid=10457.

Eisenhower, Dwight D. "The Chance for Peace: Address by President Eisenhower to the American Society of Newspaper Editors." April 16, 1953. In *Documents on International Affairs, 1953*. Royal Institute of International Affairs. Edited by Denise Folliot. London: Oxford University Press, 1956.

Eisenhower, Dwight D. "Farewell Address." January 17, 1961. http://www.eisenhower.archives.gov/research/online_documents/farewell_address/Reading_Copy.pdf.

Eisenhower, Dwight D. "The United States never lost a soldier. . . ." Quoted in Peter G. Lyon, *Eisenhower: Portrait of the Hero*. Boston: Little, Brown, 1974.

Eisenhower, Susan. *Mrs. Ike: Memories and Reflections on the Life of Mamie Eisenhower*. Sterling, VA: Capital Books, 2002. First published 1996.

Fleegler, E. W. "Firearm legislation and firearm-related fatalities in the United States." *JAMA Internal Medicine* 173 (9) (2013): 732–740.

Hoyert, Donna L., and Jiaquan Xu. "Deaths: Preliminary Data for 2011." https://www.cdc.gov/nchs/data/nvsr/nvsr61/nvsr61_06.pdf.

LaSalle, Mick. "'RoboCop' Review: No Mere Remake, a Timely Upgrade." *San Francisco Chronicle*, February 11, 2014. http://www.sfgate.com/movies/article/RoboCop-review-no-mere-remake-a-timely-upgrade-5224724.php.

Metropolitan Museum of Art. Citation of Dwight D. Eisenhower as Life Fellow, April 2, 1946. https://www.monumentsmenfoundation.org/discoveries/eisenhower-at-the-met

Metropolitan Museum of Art. "Transcript of Dwight D. Eisenhower's Address at the Metropolitan Museum on April 2, 1946." http://www.metmuseum.org/blogs/now-at-the-met/features/2011/this-weekend-in-met-history-april-2.

New York Times. "1950 Class Day Celebrated on Morningside Heights." In "Hoffman Finds U.S. Better than 1900," June 8, 1950.

Newton, Jim. *Eisenhower: The White House Years*. New York: Anchor Books, 2011.

Pietrusza, David. *1948: Harry Truman's Improbable Victory and the Year that Transformed America*. New York: Union Square Press, 2011.

Schwartz, Jason L., and Arthur L. Caplan, eds. *Vaccination Ethics and Policy: An Introduction with Readings*. Cambridge: MIT Press, 2017.

Sloan, J. H., et al. "Handgun regulations, crime, assaults, and homicide. A tale of two cities." *New England Journal of Medicine* 319 (1988): 1256–62.

Wall Street Journal. "Transcript of President's News Conference." August 21, 1958.

Wiebe, D. J., et al. "Homicide and geographic access to gun dealers in the United States." *BMC Public Health* 9 (2009): 199. doi:10.1186/1471-2458-9-199.

5. Nobel on Columbus Avenue

Beecher, Henry K., and Mark D. Altschule. *Medicine at Harvard: The First 300 Years*. Hanover, NH: University Press of New England, 1977.

Castle, William. "The Conquest of Pernicious Anemia." In *Blood, Pure and Eloquent*. Edited by Maxwell M. Wintrobe. New York: McGraw-Hill, 1980.

Fire, A. Z., et al. "Potent and specific genetic interference by double-stranded RNA in *Caenorhabditis elegans*." *Nature* 391 (1998): 806–81.

Fire, Andrew Z. "Nobel Lecture: Gene Silencing by Double Stranded RNA." December 8, 2006. Nobelprize.org. https://www.nobelprize.org/nobel_prizes/medicine/laureates/2006/fire-lecture.html.

Hooke, Robert. *Preface to Micrographia*. New York: Dover, 1962. First published 1665.

Mello, Craig C., and Adam Smith. "Interview with Craig C. Mello." Transcript of the telephone interview with Professor Craig C. Mello immediately following the announcement of the 2006 Nobel Prize in Physiology or Medicine. October 2, 2006. Nobelprize.org. http://www.nobelprize.org/nobel_prizes/medicine/laureates/2006/mello-telephone.html

Minot, G. R., and W. P. Murphy. "Observations on patients with pernicious anemia partaking of a special diet. A clinical aspect." *Transactions of the Association of American Physicians* 41 (1926): 72–5.

Napoli, C., et al. "Introduction of a chimeric chalcone synthase gene into petunia results in reversible co-suppression of homologous genes in trans." *Plant Cell* 2 (1990): 279–289.

Nobelprize.org. "The Nobel Prize in Physiology or Medicine 1934." Accessed June 26, 2017. http://www.nobel.se/medicine/laureates/1934/

Sen, G. L., and H. M. Blau. "A brief history of RNAi: the silence of the genes." *FASEB Journal* 20 (2006): 1293–9.

Whipple, G. S., and F. S. Robscheit-Robbins. "Blood regeneration in severe anemia. Favorable influence of liver, heart and skeletal muscle in diet." *American Journal of Physiology* 78 (1925): 408–418.

6. Lupus and the Course of Empire

Allardyce, Jason. "Salmond Names Scots' Independence Day." *Sunday Times* (London), November 24, 2013.

Bennett, Alan. *The Madness of King George III.* London, Faber & Faber, 1992.

Bentley, E. C. "George III." In *The Complete Clerihews*. Oxford: Oxford University Press, 1981. First published 1929.

Boros, C. A., et al. "Hydrocephalus and macrocephaly: new manifestations of neonatal lupus erythematosus." *Arthritis & Rheumatology* 57 (2007): 261–266.

The British Monarchy website. "George III (r. 1760–1820)." http://www.royal.gov.uk/HistoryoftheMonarchy/KingsandQueensoftheUnitedKingdom/TheHanoverians/GeorgeIII.aspx.

Britroyals.com. "House of Stuart: Queen Mary II." http://www.britroyals.com/stuart.asp?id=mary2

Emson, H. E. "For the Want of an Heir: The Obstetrical History of Queen Anne." *BMJ* 304 (1992): 1365–1366.

George III. "Letter on the loss of America written in the 1780s." From the British Monarchy website. https://www.royal.uk/sites/default/files/media/georgeiii.pdf.

Giordano, L. "Old Jewish House Gives Unique Feel to History." *Poughkeepsie Journal* September 13, 2006.

Hift, R. J., et al. "A review of the clinical presentation, natural history and inheritance of variegate porphyria: its implausibility as the source of the 'Royal Malady.'" *Journal of Clinical Pathology* 65 (2012): 200–205.

Hughes, G. R. "Thrombosis, abortion, cerebral disease and the lupus anticoagulant." *British Medical Journal (Clinical Research Ed.)* 287 (1983): 1088–1089.

Hughes, G. R. "The cardiolipin antibody syndrome." *Clinical and Experimental Rheumatology* 3 (1985): 285–286.

Janoff, A. S., et al. "Novel liposome composition for a rapid colorimetric test for systemic lupus erythematosus." *Clinical Chemistry* 29 (1983): 1587–1592.

Kamboh, M. Ilyas, et al. "Genome-wide association study of antiphospholipid antibodies." *Autoimmune Diseases* 2013: Article ID 761046.

Lockshin, M.D., et al. "Prediction of adverse pregnancy outcome by the presence of lupus anticoagulant, but not anticardiolipin antibody, in patients with antiphospholipid antibodies." *Arthritis & Rheumatism* 64 (2012): 2311–2318.

Martin, Adam. "Trinity Church Literally Has More Money Than It Knows What to Do With." *New York* Magazine, April 25, 2013. http://nymag.com/daily/intelligencer/2013/04/trinity-church-has-more-money-than-leadership.html

National Archives online. "Declaration of Independence: A Transcription." http://www.archives.gov/exhibits/charters/declaration_transcript.html

Petri, M., et al. "Derivation and validation of the systemic lupus international collaborating clinics classification criteria for systemic lupus erythematosus." *Arthritis & Rheumatology* 64 (2012): 2677–2686.

Philips, A. "Text for George Frideric Handel's 'Ode for the Birthday of Queen Anne.'" In *The New Grove Dictionary of Music and Musicians.* Edited by Stanley Sadie and John Tyrell. London: Oxford University Press, 2004.

Rey, D. M. "Around Annapolis: City Charter Anniversary Medal Recalls Annapolis's Vibrant Past." *Capital Gazette,* September 12, 2008.

Schama, Simon. *A History of Britain, Vol. 2: The Wars of the British.* New York: Hyperion, 2001a.

Schama, Simon. "Britannia Incorporated." Episode 10 of *A History of Britain.* London: BBC One, 2001b.

Somerset, Anne. *Queen Anne: The Politics of Passion.* New York: Knopf, 2013.

Statute of Anne, 1710. http://archive.org/stream/thestatuteofanne33333gut/33333.txt.

Trevelyan, G. M. *A Shortened History of England.* New York: Penguin Books, 1987.

Walpole, Horace. *Journal of the Reign of King George the Third: from the Year 1771 to 1783.* Vol. 2. London: Richard Bentley, 1859. https://archive.org/details/journalreignkin00johgoog

Weir, Alison. *Britain's Royal Families: The Complete Genealogy.* London: Bodley Head, 1999.

Westminster-Abbey.org. "Frederick Louis, Prince of Wales." http://www.westminster-abbey.org/our-history/people/frederick-louis,-prince-of-wales

7. Groucho on the Gridiron

ASC Faculty Leadership Committee. "Arts and Sciences Education at The Ohio State University: A Statement of Principles with Respect to the Undergraduate Curriculum." Ohio State University online. Accessed January 17, 2015. https://artsandsciences.osu.edu/sites/artsandsciences.osu.edu/files/statement_of_principles_undergradcurr.pdf.

Ballotpedia. "Governor of Ohio." Accessed January 17, 2015. http://ballotpedia.org/Governor_of_Ohio.

Bovie, Smith Palmer. *The Satires and Epistles of Horace*. Chicago: Chicago University Press, 1959.

Chronicle of Higher Education online. "Chronicle Data." Accessed January 17, 2015. http://chronicle.com/article/2013-14-AAUP-Faculty-Salary/145679?cid=megamenu.

Danzig, Allison. *The History of American Football: Its Great Teams, Players, and Coaches*. New York: Prentice-Hall, 1956.

Detroit Free Press. "Hugo Bezdek Was No Professional 'Pug' After All." December 11, 1932, p. 42.

Dombrowski, D. A. "Plato and athletics." *Journal of Physical Education and Sport* 6 (1979): 29–38.

Dupont, Kevin Paul. "MIT's Brilliance on Football Field Leads to Undefeated Season." *Boston Globe*, November 21, 2014.

Eliot, C. W. "The New Education." *Atlantic Monthly* 23 (1869): 365–366.

England, E. B. *Plato, "The Laws."* London: Longmans, Green & Co., 1921.

Forbes online. "College Football's Most Valuable Teams." Accessed January 17, 2015. http://www.forbes.com/pictures/emdm45el/ohio-state-university-buckeyes/.

Huxley, T. H. "A Liberal Education and Where to Find It." In *Science & Education*. Vol. 3 of *Collected Essays*. New York: D. Appleton, 1897. First published 1868.

Keyser, Jayson. "Robert Morris University Becomes First to Recognize Video Games As Varsity Sport." *Associated Press Wire,* October 6, 2014.

Neils, Jenifer. "The Panathenaea: an Introduction." In *Goddess and Polis: The Panathenaic Festival in Ancient Athens*. Princeton, NJ: Princeton University Press, 1992.

Ohio State University website. Accessed January 17, 2015. http://osu.edu.

Pegler, W. "Nobody's Business." *Chicago Daily Tribune*, December 11, 1932.

Perelman, S. J., et al. *Horse Feathers*. Directed by Norman Z. McLeod. Paramount Pictures, 1932.

Revsine, Dave. *The Opening Kickoff: The Tumultuous Birth of a Football Nation*. Guilford, CT: Pequot Press, 2014.

Rhoden, William C. "Quarterback Is a Work in Progress, but Oh, What Progress." *New York Times*, January 13, 2015.

Savage, Howard J., et al. *American College Athletics*. New York: Carnegie Foundation for the Advancement of Teaching, 1929.

Sports Illustrated online. "NFL's Average Salaries by Position." Accessed January 17, 2015. http://www.si.com/nfl/photos/2013/01/31/nfl-average-salaries-position#1.

Strauss, Ben, and Steve Eder. "Labor Board to Review Northwestern Football Case." *New York Times*, April 24, 2014.

Tracy, Marc, and Tim Rohan. "What Made College Football More Like the Pros? $7.3 Billion, for a Start." *New York Times*, December 30, 2014.

Tracy, Marc. "Top Conferences to Allow Aid for Athletes' Full Bills." *New York Times*, January 17, 2015.

8. Apply Directly to Forehead: Holmes, Zola, and Hennapecia

Abdulla, K. A., and N. M. Davidson. "A woman who collapsed after painting her soles." *Lancet* 348 (1996): 658.

Arck, P. C., et al. "Towards a 'free radical theory of graying': melanocyte apoptosis in the aging human hair follicle is an indicator of oxidative stress induced tissue damage." *FASEB Journal* 20 (2006): 1567–1569.

Babich, H., and A. Stern. "In vitro cytotoxicities of 1,4-naphthoquinone and hydroxylated 1,4-naphthoquinones to replicating cells." *Journal of Applied Toxicology* 13 (1993): 353–358.

Berth-Jones, J., and P. E. Hutchinson. "Novel cycle changes in scalp hair are caused by etretinate therapy." *British Journal of Dermatology* 132 (1995): 367–375.

Blair, J., et al. "Dermatitis associated with henna tattoo. 'safe' alternative to permanent tattoos carries risk." *Postgraduate Medicine* 116 (2004): 63–65.

Center for Food Safety and Applied Nutrition, Office of Cosmetics and Colors. "Temporary Tattoos and Henna/Mehndi: Fact Sheet." U.S. Food and Drug Administration online. Accessed January 2007. http://www.cfsan.fda.gov/~dms/cos-tatt.html.

Hardwicke, J., and S. Azad. "Temporary henna tattooing in siblings—an unusual chemical burn." *Burns* 32 (2006): 658.

Holmes, Oliver Wendell. "The Young and the Old Practitioner." In *Medical Essays*. Boston: Riverside Press, 1871.

Jung, P., et al. "The extent of black henna tattoos' complications are not restricted to PPD-sensitization." *Contact Dermatitis* 55 (2006): 57–59.

Kang, I. J., and M. H. Lee. "Quantification of para-phenylenediamine and heavy metals in henna dye." *Contact Dermatitis* 55 (2006): 26–29.

Kök, A. N., et al. "Henna (Lawsonia inermis Linn.) induced haemolytic anaemia in siblings." *International Journal of Clinical Practice* 58 (2004): 530–32.

Laumann, A. E., and A. J. Derick. "Tattoos and body piercings in the United States: a national data set." *Journal of the American Academy of Dermatology* 55 (2006): 413–421.

Markos, Kibret. "Scarred Children: Two North Jersey Families Are Suing Over Permanent Marks Their Kids Got from Temporary Tattoos." *Bergen County Record*, January 5, 2007.

McMillan, D. C., et al. "Role of oxidant stress in lawsone-induced hemolytic anemia." *Toxicological Sciences* 82 (2004): 647–655.

Nohynek, G. J., et al. "Toxicity and human health risk of hair dyes." *Food and Chemical Toxicology* 42 (2004): 517–543.

Onder, M., "Temporary holiday 'tattoos' may cause lifelong allergic contact dermatitis when henna is mixed with PPD." *Journal of Cosmetic Dermatology* 3 (2003): 126–130.

Price, V. H. "Androgenetic alopecia in women." *Journal of Investigative Dermatology Symposium Proceedings* 8 (2003): 24–27.

Starr, J. C., et al. "Immediate type i asthmatic response to henna following occupational exposure in hairdressers." *Annals of Allergy, Asthma & Immunology* 48 (1982): 98–99.

Thomas, L., et al. "Prevention by cortisone of the changes in cartilage induced by an excess of vitamin A in rabbits." *American Journal of Pathology* 42 (1963): 271–283.

Uygur-Bayramicli, O.et al. "Acute chemical colitis resulting from oral intake of henna." *Journal of Clinical Gastroenterology* 39 (2005): 920–921.

Zola, Émile. *The Experimental Novel, and Other Essays.* Translated by Belle M. Sherman. New York: Haskell House, 1964.

Zola, Émile. *Nana.* Project Gutenberg, accessed January 2007. http://www.gutenberg.org/etext/5250.

Science Fictions

9. Swift-Boating Darwin: Alternative and Complementary Science

Changeux, Jean-Pierre. *La lumière au siècle des lumières et aujourd'hui.* Paris: Odile Jacob, 2005.

Columbia University Medical Center. "Rosenthal Center Celebrates Tenth Anniversary." *In Vivo* 2, no. 21, December 22, 2003. http://www.cumc.columbia.edu/publications/in-vivo/Vol2_Iss21_dec22_03/around_and_about.html.

Crowther, R. Letter to the editor. *New York Times,* December 23, 2005.

Darnton, Robert. *Mesmerism and the End of the Enlightenment in France.* Cambridge: Harvard University Press, 1968.

Darnton, Robert. "Franz Anton Mesmer." *Dictionary of Scientific Biography* 9 (1974): 325–8.

Dobzhansky, T. "Nothing in biology makes sense except in the light of evolution." *The American Biology Teacher* 35 (1973): 125–129.

Frank, Melvin, and Norman Panama. *The Court Jester.* Los Angeles: Dena Enterprises, 1956.

Goodstein, Laurie. "Issuing Rebuke, Judge Rejects Teaching of Intelligent Design." *New York Times,* December 21, 2005.

HighWire. http://highwire.stanford.edu.

Holmes, Oliver Wendell. *Medical Essays.* Reprint of the 1891 edition, Project Gutenberg, 2006. http://www.gutenberg.org/files/2700/2700-h/2700-h.htm.

Horwitz, Randy, and Daniel Muller, eds. *Integrative Rheumatology.* New York: Oxford University Press, 2011.

Huxley, Thomas Henry. *The Life and Letters of Thomas H. Huxley.* Edited by Leonard Huxley. New York: D. Appleton, 1901.

Kristof, Nicholas D. "The Hubris of the Humanities." *New York Times,* December 6, 2005.

Polkinghorne, J. "Intelligent design, creationism and its critics. philosophical, theological and scientific perspectives." *Journal of Theological Studies* 54: 460–461.

Powell, Michael. "Judge Rules Against 'Intelligent Design.'" *Washington Post*, December 25, 2005.

Thomas More Law Center. "Revolution in Evolution Is Underway, Says Thomas More Law Center." News release, January 18, 2005. http://releases.usnewswire.com/GetRelease.asp?id=41768A.

Zehr, Mary Ann. "School of Faith." *Education Week*, December 7, 2005. http://www.edweek.org/ew/articles/2005/12/07/14evangelical.h25.html?q.s=school+of+faith+2005.

U.S. Bureau of Labor Statistics. "Occupational Employment Statistics: May 2016 Occupation Profiles." https://www.bls.gov/oes/current/oes_stru.htm.

10. Spinal Irritation and the Failure of Nerve

Baker, Livia. *The Justice from Beacon Hill: The Life and Times of Oliver Wendell Holmes*. New York: HarperCollins, 1991.

Bowen, Catherine Drinker. *Yankee from Olympus: Justice Holmes and His Family*. Boston: Little, Brown, 1944.

Brooks, van Wyck. *New England: Indian Summer*. New York: E.P. Dutton, 1940.

Holmes, Oliver Wendell. *The Autocrat of the Breakfast-Table.* Boston: Phillips, Sampson, 1858.

11. Galton's Prayer

Benson, H., et al. "Study of the Therapeutic Effects of Intercessory Prayer (STEP) in cardiac bypass patients: a multicenter randomized trial of uncertainty and certainty of receiving intercessory prayer." *American Heart Journal* 151 (2006): 934–42.

Burkeman, Oliver. "If you want to get better—don't say a little prayer." *The Guardian,* April 1, 2006.

Carey, Benedict. "Long-Awaited Medical Study Questions the Power of Prayer" *New York Times,* March 31, 2006.

Darwin, Charles. *More Letters of Charles Darwin*. Vol. 2. Project Gutenberg, 2008. http://www.gutenberg.org/files/2740/2740-h/2740-h.htm.

Dusek, J. A., et al. "Study of the Therapeutic Effects of Intercessory Prayer (STEP): study design and research methods." *American Heart Journal* 143 (2002): 577–84.

Galton, Francis. "Statistical inquiries into the efficacy of prayer." *Fortnightly Review* 12 (1872): 125–35.

Gardner, Martin. "Notes of a Fringe-Watcher." *Skeptical Inquirer* 25.2 (2001). http://www.csicop.org/si/show/notes_of_a_fringe-watcher_distant_healing_and_elisabeth_targ.

Garrison, H. H., and R. E. Palazzo. "What's happening to new investigators?" *FASEB. Journal* 20 (2006): 1288–89.

Krucoff, M. W., et al. "From efficacy to safety concerns: a STEP forward or a step back for clinical research and intercessory prayer? The Study of Therapeutic Effects of Intercessory Prayer (STEP)." *American Heart Journal* 151 (2006): 762–4.

Light, M. "Prayer studies a waste of money." *Buffalo News*, April 23, 2006.

Lindsey, Albert. "Area Residents Challenging Prayer Study." *Richmond Times-Dispatch*, April 9, 2006.

Melady, Mark. "'God Factor' Defended; Prayer Study Flawed." *Worcester Telegram & Gazette*, April 13, 2006.

National Institutes of Health. "Funding Patterns." https://gsspubssl.nci.nih.gov/blog/articles?funding_patterns/201699

Schlitz, M. "Meditation, Prayer and Spiritual Healing: The Evidence." *Permanente Journal* 9 (2005): 63–66.

Targ, E. F., and E. G. Levine. "The efficacy of a mind-body-spirit group for women with breast cancer: a randomized controlled trial." *General Hospital Psychiatry* 24 (2002): 238–48.

12. Dr. Doyle and the Case of the Guilty Gene

Bell, Joseph. "The Adventures of Sherlock Holmes." *Bookman*, December 1892: 50-54.

Carr, John Dickson. *The Life of Sir Arthur Conan Doyle*. New York: Carroll & Graf, 1987.

Cuvier, Georges. *Leçons d'antatomie comparée.* 8 vols. Paris: Crochard et cie, 1800.

Holmes, Oliver Wendell. *Collected Works.* 13 vols. Boston: Houghton Mifflin, 1892.

Poe, Edgar Allan. *The Complete Poems and Stories with Selected Critical Writings.* New York: Knopf, 1946.

Shama, Simon. *Dead Certainties.* New York: Knopf, 1991.

Tilton, Eleanor M. *Amiable Autocrat: A Biography of Dr. Oliver Wendell Holmes.* New York: Henry Schuman, 1947.

Wilson, Edmund. *The Forties.* Edited by Leon Edel. New York: Farrar, Straus, Giroux, 1983.

Two for the Road

13. Swift-Boating "America the Beautiful": Katharine Lee Bates and a Boston Marriage

Adams, Henry. *The Education of Henry Adams.* Edited by Henry Cabot Lodge. New York: Penguin Classics, 1995.

Archer, Linda. Reader comment on Kim Hart, "Mitt Romney Stumps in Local Tech Circle." *Washington Post* online, April 18, 2007. http://voices.washingtonpost.com/posttech/2007/04/mitt_romney_stumps_in_local_te_1.html.

Bancroft, Hubert Howe. "Woman's Department." In *The Book of the Fair: An Historical and Descriptive Presentation of the World's Science, Art, and Industry, as Viewed through the Columbian Exposition at Chicago in 1893.* Chicago and San Francisco: The Bancroft Company, 1893. http://columbus.gl.iit.edu/.

Bates, Katherine Lee. "America the Beautiful," lyrics and recollections. Falmouth Historical Society Archives, accessed July 14, 2017. http://museumsonthegreen.org/collections/archives/

Bates, Katharine Lee. *Yellow Clover: A Book of Remembrance.* New York: E. P. Dutton, 1922.

Bates, Katharine Lee. *Selected Poems of Katharine Lee Bates.* Boston & New York: Houghton Mifflin, 1930.

Bordin, Ruth. *Alice Freeman Palmer: The Evolution of a New Woman*. Ann Arbor: University of Michigan Press, 1993.

Cleghorn, Sarah Norcliffe. *Portraits and Protests*. New York: Henry Holt, 1917.

Cleveland, Grover. "August 8, 1893: Special Session Message." https://millercenter.org/the-presidency/presidential-speeches/august-8-1893-special-session-message.

Cosner, Shaaron, and Jennifer Scanlon. *American Women Historians*. Westport, CT: Greenwood Press, 1996.

Knight, Louise W. *Citizen: Jane Addams and the Struggle for Democracy*. Chicago: University of Chicago Press, 2005.

Lorentzen, Amy. "Iowa Gay Marriage Ruling Stirs 2008 Race." Associated Press, September 1, 2007.

Richards, Laura E., and Maud Howe Elliot. *Julia Ward Howe, 1819–1910*. Boston and New York: Houghton Mifflin, 1915.

Schwarz, Judith. "'Yellow Clover': Katharine Lee Bates and Katherine [*sic*] Coman." *Frontiers* 4 (1979): 59–67.

14. Alice James and Rheumatic Gout

Burr, Anna Robeson, ed. *Alice James: Her Brothers–Her Journal*. New York: Dodd, Mead, 1934.

Darwin, Charles. *The Life and Letters of Charles Darwin*. Edited by Francis Darwin. New York: Basic Books, 1959.

Desmond, Adrian, and James Moore. *Darwin: The Life of a Tormented Evolutionist*. New York: Warner, 1991.

Garrod, Alfred Baring. "Observations on certain pathological conditions of the blood and urine in gout, rheumatism and Bright's disease." *Medico-Chirurgical Transactions* 31 (1848): 83–97.

Garrod, Alfred Baring. *A Treatise on Gout and Rheumatic Gout (Rheumatoid Arthritis),* 3rd ed. London: Longmans, Green, 1876.

Garrod, Alfred Baring. "Aix-les-Bains: the value of its course in rheumatoid arthritis, gout, rheumatism, and other diseases." *Lancet* 1 (1889): 869–71.

Garrod, Alfred Baring. "Inborn errors of metabolism." *Lancet* 2 (1908): 1–7, 73–79, 142–148, 214–220.

Lagier, R. "Nosology versus pathology, two approaches to rheumatic diseases illustrated by Alfred Baring Garrod and Jean-Martin Charcot." *Rheumatology* 40 (2001): 467–7.

Matthiessen, F. O. *The James Family: Including Selections from the Writings of Henry James, Senior, William, Henry, and Alice James*. New York: Knopf, 1961.

Pardee, A., F. Jacob, and J. Monod. "The genetic control and cytoplasmic expression of 'inducibility' in the synthesis of ß-galactosidase by *E. coli*." *Journal of Molecular Biology* 1 (1959): 165.

Porter, Roy, and G. S. Rousseau. *Gout: The Patrician Malady*. New Haven: Yale University Press, 1998.

Rodnan, G. P., et al. "Sir Alfred Baring Garrod, FRS." *JAMA* 224 (1973): 663–5.

Schine, Cathleen. *Alice in Bed*. New York: Knopf, 1983.

Shorter, Edward. *From Paralysis to Fatigue: A History of Psychosomatic Illness in the Modern Era*. New York: Free Press, 1993.

Showalter, Elaine. *Hysteries*. New York: Columbia University Press, 1998.

Sontag, Susan. *Alice in Bed*. New York: Farrar, Straus and Giroux, 1993.

Strouse, Jean. *Alice James: A Biography*. Boston: Houghton Mifflin, 1980.

Weissmann, G., and G. Rita. "The Molecular basis of gouty inflammation: interaction of monosodium urate crystals with lysosomes and liposomes." *Nature New Biology* 240 (1972): 167–172.

Wessely, S. "Medically unexplained symptoms: exacerbating factors in the doctor-patient encounter." *Journal of the Royal Society of Medicine* 96 (2003): 223–7.

Wessely, S. and M. Hotopf. "Is fibromyalgia a distinct clinical entity? Historical and epidemiological evidence," *Best Practice & Research: Clinical Rheumatology* 13 (1999): 427–36.

15. Free Radicals Can Kill You: Lavoisier and the Oxygen Revolution

Briggs, R. T., et al. "Localization of NADH oxidase on the surface of human polymorphonuclear leukocytes by a new cytochemical method." *Journal of Cell Biology* 67 (1975): 566–586.

Cohen, I. Bernard. *Revolution in Science*. Cambridge: Belknap, Harvard University Press, 1985.

Djerassi, Carl, and Roald Hoffmann. *Oxygen*. Weinheim, Germany: Wiley-VCH Verlag, 2001.

Eagle, C. T., and J. Sloan. "Marie Anne Paulze Lavoisier: the mother of modern chemistry." *Journal of Chemical Education* 3 (1998): 1–18.

FASEB Journal online. Accessed January 25, 2010. http://www.fasebj.org/search.dtl.

Gerschman, R., et al. "Oxygen poisoning and x-irradiation: a mechanism in common." *Science* 119 (1954): 623–626.

Goldstein, I. M., et al. "Ceruloplasmin. A scavenger of superoxide anion radicals." *Journal of Biological Chemistry.* 254 (1979): 4040–4045.

Grimaux, Edouard. *Lavoisier, 1743–1794, d'après sa correspondance, ses manuscrits, ses papiers*. Paris: Cuchet, 1896.

Guerlac, Henry. *Antoine-Laurent Lavoisier, Chemist and Revolutionary*. New York: Charles Scribner's, 1975.

Harman, D. "Aging: a theory based on free radical and radiation chemistry." *Journals of Gerentology.* July 11, 1956: 298–300.

Harman, D. "Origin and evolution of the free radical theory of aging: a brief personal history, 1954–2009." *Biogerontology* 10 (2009): 773–781.

Hartley, H. "Antoine Laurent Lavoisier." *Proceedings of the Royal Society of London B: Biological Sciences* 134 (1947): 348–377.

Hoffmann, R. "Mme Lavoisier." *American Scientist* 90 (2002): 22–24.

Hutchins, Robert Maynard, ed. *Lavoisier, Fourier, Faraday*. Vol. 45 of *Great Books of the Western World.* Encyclopædia Britannica, 1952.

Lavoisier, Antoine-Laurent. "*Mémoire sur la nature du principe qui se combine avec les métaux pendant leur calcination et qui en augmente le poids.*" In *Histoire de l'Académie Royale des Sciences, 1775*. Paris: L'imprimerie Royale, 1778.

Lavoisier, Antoine-Laurent. *Traité Élémentaire de Chimie*. Paris: Cuchet, 1789.

McCord, J. M., and I. Fridovich. "The utility of superoxide dismutase in studying free radical reactions. I. Radicals generated by the interaction of sulfite, dimethyl sulfoxide, and oxygen." *Journal of Biological Chemistry* 244 (1969): 6056–6063.

McKie, Douglas. *Antoine Lavoisier: Scientist, Economist, Social Reformer*. New York: Schumann, 1952.

Pinault-Sörensen, Madeleine. "*Madame Lavoisier, dessinatrice et peintre.*" *La Revue du musée des arts et métiers*, March 1994: 23–25.

Priestley, Joseph. *Experiments and Observations on Different Kinds of Air*. Vol. 2. Birmingham, AL: Thomas Pearson, 1775.

Scheele, Carl Wilhelm. *Chemische Abhandlung von der Luft und dem Feuer. Nebst einem Vorbericht von Torbern Bergman*. Uppsala and Leipzig: Magnus Swederus, Johan Edman & Siegfried Lebrecht Crusius, 1777.

Severinghaus, J. W. "Fire-air and dephlogistication. revisionisms of oxygen's discovery." *Advances in Experimental Medicine and Biology* 543 (2003): 7–19.

Van Klooster, H. S. "Franklin and Lavoisier." *Journal of Chemical Education* 23 (1946): 107–109.

16. Dr. Blackwell Returns from London

Altman, Lawrence K. "Agency Urges a Change in Antibiotics for Gonorrhea." *New York Times*, April 13, 2007.

Baker, S. Josephine. *Fighting for Life*. New York: New York Review Books, 2013. First published 1939.

Bates, Katharine Lee. *Selected Poems of Katharine Lee Bates*. Boston & New York: Houghton Mifflin, 1930.

Blackwell, Elizabeth. "Ship Fever: An Inaugural Thesis. Submitted for the degree of M.D., at Geneva Medical College." *Buffalo Medical Journal and Monthly Review* 4 (1849): 523-531.

Blackwell, Elizabeth. *Pioneer Work in Opening the Medical Profession to Women*. New York: Humanity Books, 2005. First published 1865.

Blot, Hippolyte. *De l'albuminurie chez les femmes enceintes: ses rapports avec l'éclampsie, son influence sur l'hémorrhagie utérine après l'accouchement.* Paris: Rignoux, 1849.

Demirjian, Karoun. "Religious Leaders Rip Hate-Crime Measure; House Passes Bill Extending Coverage to Sexual Orientation." *Chicago Tribune*, May 5, 2007.

Elliott, H. B. "Woman as Physician." In *Eminent Women of the Age; Being Narratives of the Lives and Deeds of the Most Prominent Women of the Present Generation.* Edited by James Parton et al. Hartford: S. M. Betts & Co., 1868.

Holmes, Oliver Wendell. *Currents and Counter-Currents in Medical Science.* Boston: Ticknor and Fields, 1861.

Leslie's Illustrated Newspaper. "Women's Medical College of New York Infirmary for Women and Children." April 16, 1876.

New York Times. "Bellevue Hospital Medical College." March 1, 1867.

New York Times. "Transcript: The Republican Presidential Candidates Debate." May 3, 2007. http://www.nytimes.com/2007/05/03/us/politics/04transcript.html?ex=133601.

Punch. "An M.D. in a Gown." Vol. 16 (January–June 1849): 226.

Shrier, D. "A celebration of women in US medicine." *Lancet* 363 (2004): 253-253.

Tocqueville, Alexis de. *Recollections: The French Revolution of 1848.* Edited by J. P. Mayer and A. P. Kerr. New Brunswick and London: Transaction Publishers, 2009.

17. Call Me Madame

Bensaude-Vincent, Bernadette. *Langevin, 1872–1946: Science et Vigilance.* Paris: Belin, 1987.

Curie, Ève. *Madame Curie: A Biography.* Translated by Vincent Sheehan. New York: Da Capo Press, 1986. First published 1937.

Curie, Marie. *Correspondance: choix de lettres, 1905–1934 [de] Marie [et] Irène Curie.* Paris: Éditeurs français réunis, 1974.

Giroud, Françoise. *Marie Curie, a Life.* Translated by Lydia Davis. New York: Holmes & Meier, 1986.

Igot, Y. *Monsieur et Madame Curie.* Paris: Didier, 1960.

Langevin, Paul. *La pensée et l'action.* Edited by Paul Labérenne. Paris: Editions sociales, 1964.

Pais, Abraham. *Subtle Is the Lord: The Science and the Life of Albert Einstein.* Oxford: Oxford University Press, 1982.

Pais, Abraham. *A Tale of Two Continents: The Life of a Physicist in a Turbulent World.* Princeton: Princeton University Press, 1997.

Quinn, Susan. *Marie Curie: A Life.* New York: Simon & Schuster, 1995.

Woolf, Harry, ed. *Some Strangeness in the Proportion: A Centennial Symposium to Celebrate the Achievements of Albert Einstein.* Reading, MA: Addison-Wesley, 1980.

Beside the Golden Door

18. Welcome to America: Einstein's Letter to the Dean

Anonymous (n.d.) Bekannte Altenstädter: Rudolf Ehrmann der Altenstädter, der zu den 50 berühmtesten Deutschen gehörte. Available at http://www.altenstadt.de/gv_altenstadt/Tourismus%20&%20Kultur/Bekannte%20Altenst%C3%A4dter/Bekannte%20Altenst%C3%A4dter%204.pdf

Block, S. R. "Currier McEwen MD, 1902–2003." *Arthritis & Rheumatology* 48 (2003): 2739–2740.

Breitman, Richard, and Alan M. Kraut. *American Refugee Policy and European Jewry, 1933–1945.* Bloomington, IN: Indiana University Press, 1987.

Cannon, Walter B. *The Way of an Investigator: A Scientist's Experiences in Medical Research.* New York: Hafner Publishing Co., 1945.

Carson, S. L. "The conspiracy between Einstein and my doctor against the Nazis . . . and the U.S." *Manuscripts* 45 (1989): 265–267.

Centers for Disease Control and Prevention. "Minority Health: Asian American Populations." Accessed October 6, 2015. http://www.cdc.gov/minorityhealth/populations/REMP/asian.htm.

Einstein, Albert. "Letter to Franklin D. Roosevelt, August 2, 1939." http://www.dannen.com/ae-fdr.html.

Ehrmann, R. "The effect of adrenaline on the skin glandular secretion of the frog." *Archiv für Experimentelle Pathologie und Pharmakologie* 53 (1905): 137–139.

Ehrmann, R. "Ein Gefaessprozess bei Lues." *Wiener Medizinische Wochenschrift* 57 (1907): 777–782.

Ehrmann, R., and R. Lederer. "The conduct of the pancreas with achylia and anacidity of the stomach." *Deutsche Medizinische Wochenschrift* 35 (1909): 879–883.

Girdusky, Ryan. "72 House Democrats ask Obama to take in 100,000 Syrian refugees." *Red Alert Politics*, September 13, 2015. http://redalertpolitics.com/2015/09/13/72-house-democrats-ask-obama-take-100000-syrian-refugees/.

Grady, Denise. "Lasker Prizes Given for Discoveries in Cancer and Genetics and Ebola Response." *New York Times*, September 9, 2015.

Guardian online. "Why Border Walls—Even with Canada—Are Not the Republicans' Trump card." August 31, 2015. http://www.theguardian.com/us-news/2015/aug/31/border-walls-canada-mexico-republicans-walker-trump.

Johnson, Lyndon B. "President Lyndon B. Johnson's Remarks at the Signing of the Immigration Bill, Liberty Island, New York, October 3, 1965." http://www.lbjlib.utexas.edu/johnson/archives.hom/speeches.hom/651003.asp.

Lazarus, Emma. "The New Colossus." In *Emma Lazarus: Collected Poems.* Edited by John Hollander. New York: Literary Classics, 2005.

Long, Breckinridge. Letters to Almy Edmunds (July 1933) and to Joseph E. Davies (September 1933) (July 1933). Quoted in Fred L. Israel, ed., *The War Diary of Breckinridge Long*, p. xvii. Lincoln: University of Nebraska Press, 1966.

Long, Breckinridge. Memo to Adolf Berle, Jr., and James Dunn, June 26, 1940. Quoted in Doris Kearns Goodwin, *No Ordinary Time: Franklin & Eleanor Roosevelt; The Home Front in World War II.* New York: Simon & Shuster, 1994.

New York Times. "Dr. Albert Einstein Dies in Sleep at 76, World Mourns Loss of Great Scientist." April 18, 1955.

Vilcek Foundation. "Immigrant Nation, American Success: Achievements in STEM." Accessed October 6, 2015. http://www.vilcek.org/news/current-news/immigrant-nation-american-success-achievements-in-stem.html.

19. Modernism and the Hippocampus: Kandel's Vienna

Braka, Ivor. Quoted in Carol Vogel, "'The Scream' Is Auctioned for a Record $119.9 Million." *New York Times*, May 2, 2012.

Kandel, Eric. *The Age of Insight: The Quest to Understand the Unconscious Brain, from Vienna 1900 to the Present*. New York: Random House, 2012.

Kandel, Eric, James H. Schwartz, and Thomas Jessell. *Principles of Neural Science*. 4th ed. New York: McGraw-Hill, 2000.

20. A Taste of the Oyster: Vilcek's Love of Science

Chargaff, Erwin. *Heraclitean Fire*. New York: Rockefeller University Press, 1978.

Crick, Francis. *What Mad Pursuit: A Personal View of Scientific Discovery*. New York: Basic Books, 1988.

Knight, D. M., et al. "Construction and initial characterization of a mouse-human chimeric anti-TNF antibody." *Molecular Immunology* 30 (1993): 1443–53.

Medawar, Peter. *Memoir of a Thinking Radish*. Oxford: Oxford University Press, 1988.

Vilcek, J. "An interferon-like substance released from tick-borne encephalitis virus-infected chick embryo fibroblast cells." *Nature* 187 (1960): 73–4.

Vilcek Foundation online. "2015 Prize Recipients." Accessed January 10, 2016. http://www.vilcek.org/prizes/prize-recipients/2015/index.html.

Walter, Eugene. "Federico Fellini: Wizard of Film." *Atlantic Monthly* 216 (1965): 62–68.

21. Richard Dawkins Lights a *Brief Candle in the Dark*

Comfort, N. "Genetics: Dawkins, redux." *Nature* 525 (2015): 184–185.

Dawkins, R., and H. J. Brockmann. "Do digger wasps commit the Concorde fallacy?" *Animal Behaviour* 28 (1980): 892–896.

Dawkins, R., and J. R. Krebs. "Arms races between and within species." *Proceedings of the Royal Society of London B: Biological Sciences* 205 (1979): 489–511.

Dawkins, Richard. *The Extended Phenotype: The Long Reach of the Gene.* Oxford: Oxford University Press, 1999.

Dawkins, Richard. *The Selfish Gene: 30th Anniversary Edition.* Oxford: Oxford University Press, 2006.

Dawkins, Richard. *The God Delusion.* New York: Random House, 2009.

Dawkins, Richard. *An Appetite for Wonder: The Making of a Scientist.* New York: Random House, 2013.

Dawkins, Richard. *Brief Candle in the Dark: My Life in Science.* New York: HarperCollins, 2015.

Discovery Institute. "Creationist Defenses, 2014." *Creation Science Hall of Fame.* Accessed October 22, 2015. http://creationsciencehalloffame.org/defenses/creationist-defenses.

McRee, Patrick, dir. *The God Delusion Debate.* Birmingham, AL: Fixed Point Foundation, 2007.

22. Eugenics and the Immigrant: Rosalyn Yalow

Barrett, Craig. "Why America Needs to Open Its Doors Wide to Foreign Talent." *Financial Times,* January 31, 2006.

Dutt, Ela. "Scientists Denied U.S. Visa; Prof. Mehta, Prominent Scientist, Applied for a Visa and That Is Being Issued." *News India-Times,* March 3, 2006.

Mervis, J. A. "Glass ceiling for Asian scientists?" *Science* 310 (2005): 606–607.

Odelberg, Wilhelm, ed. "Rosalyn Yalow." In *Les Prix Nobel. The Nobel Prizes 1977.* Stockholm: Nobel Foundation, 1978. http://www.nobelprize.org/nobel_prizes/medicine/laureates/1977/yalow-bio.html.

Pearson, K., and M. Moul. "The problem of alien immigration into Great Britain, illustrated by an examination of Russian and Polish Jewish children." *Annals of Eugenics* 1 (1925): 1–128.

Levi-Montalcini, Rita. *In Praise of Imperfection: My Life and Work*. New York: Basic Books, 1988.

Thomson Scientific. "Science Watch Study Shows United States Loses Dominant Share of World Science." Press release, July 28, 2005. http://scientific.thomson.com/press/2005/8282889.

United States Census Bureau. "Industry and Occupation Data." https://www.census.gov/people/io/data/.

23. Cortisone and the Burning Cross

Chamberlain, Gaius. "Percy Julian." The Black Inventor Online Museum. Accessed June 23, 2017. http://blackinventor.com/percy-julian/.

Hargraves, M. M., and R. Morton. "Presentation of 2 bone marrow elements—the tart cell and the LE cell." *Proceedings of the Staff Meetings Mayo Clinic* 23 (2) (1948): 25–28.

Hench, P. S. "A Nobel prize story." *Patient Care* 33 (1999): 173-74.

Hench, P. S., et al. "The effect of a hormone of the adrenal cortex (17-hydroxy-11 dehydrocorticosterone: compound E) and of pituitary adrenocorticotropic hormone on rheumatoid arthritis." *Proceedings of the Staff Meetings Mayo Clinic* 24 (1949): 181–297.

Hench, Philip S. "The Reversibility of Certain Rheumatic and Non-Rheumatic Conditions by the Use of Cortisone or of the Pituitary Adrenocorticotropic Hormone." In *Nobel Lectures, Physiology or Medicine 1942–1962*. Amsterdam: Elsevier, 1964. http://www.nobelprize.org/nobel_prizes/medicine/laureates/1950/hench-lecture.pdf.

Julian, P. L., et al. "Sterols XI. 17α-hydroxy-11-desoxycorticosterone (Reichstein Substance-S)." *Journal of the American Chemical Society* 72 (11) (1950): 5145–5147.

Laurence, William A. "Aid in Rheumatoid Arthritis Is Promised by New Hormone." *New York Times*, April 21, 1949.

New York Times. "Arson Fails at Home of a Negro Scientist." November 23, 1950.

Nobelprize.org. "The Nobel Prize in Physiology or Medicine 1950." Nobel Media A B 2014. http://www.nobel.se/medicine/laureates/1950.

Rose, H. M., et al. "Differential agglutination of normal and sensitized sheep erythrocytes by sera of patients with Rheumatoid Arthritis." *Proceedings of the Society for Experimental Biology and Medicine* 68 (1) (1948): 1–6.

Sarrett, L. H. "The partial synthesis of dehydrocorticosterone acetate." *Journal of the American Chemical Society* 68 (1946): 2478–2483.

Witkop, Bernhard. *Percy Lavon Julian, 1899–1975: A Biographical Memoir.* Vol. 52 of *Biographical Memoirs.* Washington, D.C.: National Academy Press, 1980. http://www.nasonline.org/publications/biographical-memoirs/memoir-pdfs/julian-percy.pdf.

Ave atque Vale

24. Lewis Thomas and the Two Cultures

Barzun, Jacques. *A Stroll with William James.* Chicago: University of Chicago Press, 1983.

Gould, Stephen Jay. "Calling Dr. Thomas." In *An Urchin in the Storm: Essays about Books and Ideas.* New York and London: W. W. Norton, 1987.

Hook, Sidney. "How to Blow Your Own Horn Effectively." *Wall Street Journal,* November 23, 1987.

Leclerc de Buffon, Georges-Louis. *Discours sur le style: texte de l'édition de l'abbé J. Pierre.* Paris: Librairie Ch. Poussielgue, 1896.

Lehmann-Haupt, Christopher. Review of *The Fragile Species,* by Lewis Thomas. *New York Times,* April 16, 1992.

Lovelock, James. *GAIA: A New Look at Life on Earth.* Oxford: Oxford University Press, 1979.

Silverstein, Arthur M. *A History of Immunology.* New York: Academic Press, 1989.

Snow, C. P. *Two Cultures and the Scientific Revolution.* Cambridge: Cambridge University Press, 1959.

Thomas, Lewis. "Limitation." *Atlantic Monthly* 174: 119 (1944).

Thomas, Lewis. *The Lives of a Cell.* New York: Viking Press, 1974.

Thomas, Lewis. *The Youngest Science.* New York: Viking, 1983.

Thomas, Lewis. *The Fragile Species.* New York: Scribner's, 1992.

Thomas, Lewis. *Late Night Thoughts on Listening to Mahler's Ninth Symphony*. New York and London: Penguin Books, 1995.

Waugh, Evelyn. *A Little Order*. Edited by Donat Gallagher. Boston: Little, Brown & Co., 1977.

Zinsser, Hans. *As I Remember Him: The Biography of R. S.* Boston: Little, Brown and Co., 1940.

Acknowledgements

MANY OF THESE ESSAYS appeared originally in *The FASEB Journal*, the official journal of the Federation of American Societies of Experimental Biology, a publication for which I was responsible from 2006–2016. An armful of thanks for the energetic staff in Bethesda, especially Jennifer Pesanelli, Cody Mooneyhan, Mary Hayden, and Susan Moore. At the NYU School of Medicine, Ms. Andrea Cody, the administrator of the Biotechnology Study Center at the NYU School of Medicine, has been responsible for keeping my prose intelligible to humans. Above all: thanks to the Bellevue Literary Press, with its multi-talented director, Erika Goldman, and to Elana Rosenthal and Marjorie DeWitt, who slid this work into print. The BLP has made all of us at the university—and the hospital for which the press is named—very proud, indeed. The MBL/WHOI library, where I work in the summers, is not only a repository of marine wisdom but also a community of scholars, archivists, and active scientists. Writing at Woods Hole has been made a pleasure by the late Cathy Norton, by Diane Rielinger, Jennifer Walton, Matthew Person, John Furfey, and Nancy Stafford.

Index

Bellevue Literary Press is devoted to publishing literary fiction and nonfiction at the intersection of the arts and sciences because we believe that science and the humanities are natural companions for understanding the human experience. With each book we publish, our goal is to foster a rich, interdisciplinary dialogue that will forge new tools for thinking and engaging with the world.

To support our press and its mission, and for our full catalogue of published titles, please visit us at blpress.org

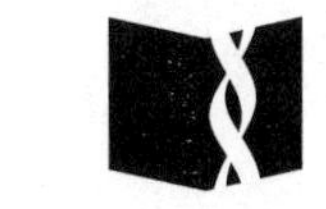

Bellevue Literary Press
New York